STUDIES IN THE HISTORY OF SCIENCE—5

General Editor: L. Pearce Williams

GENESIS OF RELATIVITY

GENESIS OF RELATIVITY

Einstein in Context

by LOYD S. SWENSON, Jr.
University of Houston

BURT FRANKLIN & CO., INC.

For
JAN, KAREN, NEIL,
and all others who
seek to understand
and
understand "to seek"

Published by Burt Franklin & Co.
235 East Forty-fourth Street
New York, New York, 10017

© 1979 by Burt Franklin & Co., Inc.
All rights reserved

Library of Congress Cataloging in Publication Data

Swenson, Loyd S
Genesis of relativity.

(Studies in the history of science; 5)
Includes bibliographical references and index.
1. Einstein, Albert, 1879–1955. 2. Relativity
(Physics) I. Title.
QC16.E5S95   530.1'1'09   79-765
ISBN 0-89102-101-9

Frontispiece photograph courtesy Lotte Jacobi

Manufactured in the United States of America

Contents

Preface

This series of studies in the history of science began with L. Pearce Williams's book *The Origins of Field Theory* (New York: Random House, 1966). That study stressed the works of Michael Faraday and James Clerk Maxwell in Great Britain during the mid-nineteenth century in creating classical electromagnetic field theory. Williams sought to create a series of studies that would "illuminate those frequent moments in the evolution of science when the scientific world finds itself torn between the old and the new." This study is my response to that challenge, a narrative sequel to Williams's book, which might have been entitled *The Origins of Relativity Theory*.

The beginnings of relativity theory may be traced deep into the past. One source goes back to Newton, Galileo, and even Zeno of Elea; another is to be found in the complicated society of western civilization in the late nineteenth and early twentieth centuries. In either case, the advent of relativistic considerations in physical science will be found to be a profoundly complex set of happenings. Whether emphasis is placed on the intellectual life internal to the history of physics or on the social and technical environments that provided the external setting for scientific thought, the focus of the relativity revolution traditionally has centered on the life and thought of one man, Albert Einstein. In 1905, he conceived the foundations of what we have learned to call the special theory of relativity. Through the next decade he developed its implications, and in 1916 he published his general theory of relativity. These theories in competition with others sought to keep abreast of momentous advances in experimental, observational, and analytical physics. The radical implications of his ideas made Einstein a hero for some, a celebrity to many, and an authority for others to rebel against. But the community of physicists studying electromagnetism worldwide was growing so large and productive that relativity theory or some equivalent appeared to many to be inevitable, at. least in its restricted or special form.

This work argues that Einstein can best be understood in the context of a follower who became a leader in the new profession of theoretical physics. My purpose is not to challenge popular notions of Einstein's greatness or his stature as a symbol of iconoclastic thinking in science. Rather, serious students of science and society are encouraged to review both the relativity revolution and Einstein's genius as cultural phenomena. Most physicists judge the quantum mechanics of the 1920s as more worthy of the political metaphor "revolution." But whereas they usually admit that quantum mechanics was heavily influenced by Weimar society, they seldom grant a social background to relativity theory. Many implications of Einstein's thought became clear to physicists and philosophers only during the period between the two world wars. Almost everyone else became aware of one such implication after the atomic bomb explosions of 1945 at Alamogordo, Hiroshima, and Nagasaki. However distorted the popular image of Einstein, relativity theory, and nuclear physics may be, people may never share an adequate understanding of them without the context provided by the history of science.

Physicists, mathematicians, philosophers, teachers, journalists, and cranks of many sorts have written around and about relativity for more than half a century. Yet there remains great interest in relativity as an esoteric part of modern science. The paradoxes of relativity theory had counterparts in the conditions of its birth and growth toward confirmation, as a few leading scholars are beginning to show us. Difficulties abound in trying to understand the relativity of simultaneity, the contraction of matter, the dilation of time, the transformation of coordinates, the gravity-inertia equivalence in relation to space-time in elementary particles and galactic astrophysics. These difficulties have led to such tantalizing logical paradoxes that most writers on relativity have given up the effort to understand the chronological and technological background of Einstein's theories.

History comprehends life better than logic does, and therefore, this work avoids the systematic mathematical logic that formed

Einstein's intellectual inheritance and environment, concentrating instead on the life he led and on the lives he loved. A decade-by-decade assessment of the process of scientific change will show that Einstein and relativity can be understood as intellectual history. My first effort to understand relativity linearly concentrated on the history of an experiment that supposedly proved false a concept that Einstein later discarded, that of an ethereal electromagnetic aether. That study of the relationships of the Michelson-Morley experiments to relativity theory led me to appreciate the role of aesthetic symmetry in pursuit of physical truth. Symmetry now appears as important as any other nonphysical value—simplicity or elegance, for example, in theoretical physics. In my earlier work I sought to re-create the supposed experimental bases for the theoretical advances of Einstein by examining the half-century life history (1880–1930) of the celebrated Michelson-Morley-Miller aether-drift tests. The pages that follow will convey primarily the roughly contemporaneous (1870–1920) theoretical success story that developed despite many experimental failures. I seek to balance the triumph of an operationally useful consensus for relativity theory with some of the tragic context of human failures, lost causes, and irrelevant issues that accompanied the relativity revolution.

The tragic dimension of science before its practitioners learned to "know sin" during and after the two world wars was often exemplified by the hybris, the overweening pride, of neo-positivistic attitudes among the new professionals. Ironically, "positivistic" philosophy rested heavily on relativistic physics for support and comfort. But the converse accusation, that relativistic physics derived largely from "positivistic" philosophy, has not withstood prolonged scrutiny. Einstein and most of his colleagues, older and younger, regardless of how specialized in interests, were themselves forced to change with the times, and science, technology, and society never before had changed so rapidly.

The theory of relativity has had epistemological consequences far beyond those envisioned at first by Einstein and his partisans. Without a fixed point or frame of reference somewhere in the cosmos, the concept of *the universe* (described by the definite article) itself becomes redundant, if not self-contradictory. Cardinal

Nicholas of Cusa (1401?–64), indeed, had defined the universe as a sphere whose center is everywhere and circumference nowhere. Age-old philosophical debates over the nature of "physical reality" and the nature of our ability "to know" had to resume their place in scientific seminars. Physics became recognized as enmeshed in metaphysics once again. Together the relativity and quantum revolutions in the 1920s led to principles of indeterminacy and complementarity so far-reaching and fundamental that aesthetic criteria such as simplicity, style, and symmetry became "scientific" values, as highly prized as the utmost standards of precision and limits of accuracy for human judgment. In short, high science became sanctified as an art form, while low science continued to feed the fires of technological transformations. In both science and engineering we are only beginning to see how psychological and sociological changes occur.

The social context of the relativity revolution developed gradually within science over a half century or more. Like most social evolution, it began slowly, then accelerated steadily in the twentieth century. My picture has been drawn in five chapters through the construction of a balanced chronological framework for the chaotic mass of historical materials available on relativity. This representation strives for verisimilitude; but the whole truth is far from accessible, and my images may at times be distorted. While other scholars analyze textual issues of proper interpretation, this book rests upon them and offers a contextual essay on the history of relativity theory. After a prologue on Einstein's stature, the chapters proceed sequentially from one great singular synthesis, that of James Clerk Maxwell for electromagnetism; into plural analyses of energy transfer and radiation phenomena; then into a decade of abortive syntheses for aether, electrons, and atoms; then into more plural analyses of relativistic invariances and equivalences; and, finally, to another great and essentially singular synthesis, that of Einstein for mass-energy, space-time, and gravity-inertia. An epilogue tells a bit more about Einstein's last quest for a unified field theory of the cosmos and about his religious views.

I hope this study in breadth will prove complementary to other studies in depth now under way by other students of relativity history. The genesis of quantum theory, a different though concurrent story, has achieved as much serious attention recently as the origins of relativity theory, but the need for reliable treatments is still unsatisfied. It will take years to set the whole story of twentieth-century physics in its proper dramatic unity. The interplay of thought and action, theory and experiment, individuals and institutions in science is both comic and tragic, despite the actors' common belief that their lines are delivered as if for a triumphal pageant rather than a tragicomic play.

Acknowledgments

This book owes much to the patience, counsel, and help of many people. Although I cannot now recall all those who over a decade have aided me in efforts to understand the historical origins of Einstein and his work, I must mention my debts to persons who have helped structure my work since the 1972 publication of *The Ethereal Aether*. L. Pearce Williams as editor and critic, Tom Franklin as publisher, Eugene Guth as a critical reader of an early draft, Otto Nathan as trustee of the estate of Albert Einstein, and Gerald Holton and Robert S. Shankland as friends and guides have been most kind. Colleagues and fellow students including Joseph A. Schatz, David H. Weinstein, Stephen G. Brush, Alfred M. Bork, Russell McCormmach, Thornton Page, Albert Van Helden, Carlo Giannoni, Peter A. Bowman, Hannah S. Decker, James B. Sullivan, Charles E. Rogers, John D. Bremsteller, Paul A. Hanle, and B. Monte Pettitt have helped me clarify many ideas. Shirley Bridgwater and my departmental staff, including Donna L. Smith, Diana G. Lucas, and Ann M. Swanzy, have done yeoman service in translating a manuscript into acceptable form. My family has also endured a long gestation period with cheerful encouragement. What follows is my responsibility but nonetheless is also, like Einstein's theories, a social product.

Our only way of avoiding the extremes of materialism and mysticism is the never ending endeavor to balance analysis and synthesis.

—Niels Bohr, 1938

Prologue on Einstein

"Relativity," as a word, evokes images of our universe in its largest and smallest aspects. Our most fundamental ideas of time and space, of matter and motion, of change and continuity, and of cause and effect are aroused whenever "relativity" is invoked. Since the advent of "atomic" bombs and nuclear power, space travel and transistor radios, there are few people anywhere who are not at least dimly aware of some sort of magic interrelationship between physical concepts like energy, mass, and the speed of light. Usually such ideas in turn call forth another image, that of a kindly old gentleman with white leonine hair and penetrating eyes, with sockless feet and a presumably fathomless mind. Widely revered as the outstanding scientist-philosopher of the twentieth century, Albert Einstein (1879–1955) is considered the creator of relativity theory. So great has been his charisma that our story can neither begin nor end without special notices of his context.

When in 1946 Einstein sat down at age sixty-seven to write "something like my own obituary" and to beg the forgiveness of Sir Isaac Newton, he recalled the beginnings of his thought on relativity theory in terms of a paradoxical wonder that had bothered him as a student at the age of sixteen in 1895: "If I pursue a beam of light with the velocity c (velocity of light in a vacuum), I should observe such a beam of light as a spatially oscillatory electromagnetic field at rest."[1]

Although Einstein was acutely aware that "today's person of 67 is by no means the same as was the one of 50, of 30, or of 21," he yet recalled his earlier childhood encounters with the mystery of a toy magnetic compass needle and, about the age of twelve, with the reputed certainty of Euclid's plane geometry. He recounted his youthful fascination with the philosophy of Ernst Mach, the mathematical thought of Henri Poincaré, and the physics of James Clerk Maxwell. These influences Einstein remembered especially

1

well as stimuli to his wonder over the question, What should I see if I could ride a wave of light?

This question, amplified into a thought-experiment so characteristic of the philosophically inclined Einstein, was conceived shortly after a telecommunications revolution had girdled the world with telegraph and telephone wires, with submarine oceanic cables, and even with wireless wave, or radio, networks in a few places. Commercial transportation systems of great complexity likewise had crisscrossed the world ocean and linked the land terminals of most major cities in regularized schedules for the exchange of goods. Steamships, railroads, electric trolleys, and elevators were making civilized men more urbanized and perhaps urbane. New thoughts were inevitable in this new world. All sophisticated Europeans could consider themselves world citizens at the turn of the twentieth century, but few dared try.

Einstein dared. After a slow start as probably a dyslexic child, he became an extraordinarily sensitive young man who wanted to understand everything, to comprehend the universe. The mysteries of magnets, electric appliances, electromagnets, and electrochemical devices, which his father for a time manufactured, presented enduring puzzles that no one could fully explain to satisfy his intellectual curiosity. He chose the physical sciences as his muse and mentor because physics seemed most fundamental to understanding the latest advances in the most basic way. As he matured within his chosen discipline, he learned that its façade hid a multitude of intractable problems. The simplicity and elegance of mathematical descriptions in certain branches of physics, especially in theoretical mechanics and thermodynamics, were by no means complete. If abstract kinematics or celestial mechanics sometimes seemed complete, thermodynamics, hydrodynamics, aerodynamics, and electrodynamics were not yet unified in a common framework. Physics was divided by separate treatments for continuous and discontinuous phenomena: there was one standard approach for ponderable matter in motion, and quite another for wave motions through seemingly empty space.

Einstein's famous paper "On the Electrodynamics of Moving Bodies" (1905) is generally credited as marking the genesis of modern relativity theory. We shall see that this is a gross oversimplification, and yet we must admit that it is somewhat more than a half-truth. Einstein's 1905 paper does contain a new way of thinking about such problems as the imaginary situation that might arise if one could surf a wave of light. For almost a decade the young Swiss patent examiner who had toyed with electromagnets had been pondering the imponderable (weightless and therefore almost unimaginable) cases of "aether in magnetic fields" and of physical conditions observable at the limits of human knowledge. Radioactivity, X rays, and other radiation phenomena together with the electron theory had opened new vistas into the realm of the ultrasmall, while astrophysics based on photography, spectroscopy, and atomic theory had recently opened new doors to the physical understanding of the ultralarge as well as the ultrasmall worlds around man. Within the previous three generations, European scientists had discovered and devoted much attention to invisible lights, inaudible sounds, unsmellable odors, intangible fields, and other phenomena transcending the unaided human senses. So many new possibilities had been uncovered that some scientists now conceived their main task to be to discover new principles of impotency. Others wished to extend indefinitely man's powers of sensory perception.

Einstein was one of those who saw through the confusion of these developments, particularly in radiation studies. He perceived a need for a new synthesis in energy-transfer studies, and he postulated the new principle of impotency that neither matter nor messages could move faster than about 300,000 km per second, the speed of light. He conceived a way to redefine extremely large or small lengths, breadths, widths, and times so that certain parameters could remain fixed despite the flux of measurements made from different viewpoints in relative motion. Indeed, in redefining optical and electromagnetic forces in terms of relative motion between bodies, Einstein raised the mechanical principle of relativity into an optical and electrical postulate. In redefining the operational meaning of the constancy of the velocity of light, he elevated that velocity from a finite value to a position virtually equivalent to an infinite value. The speed of light, c, thus became a new limit, a new absolute constant, one that represented instantaneity despite its finitude. If no material thing and no informative signal can break the optical barrier, can travel faster than the speed of light, then extremes of space and time become relative to whoever observes changes within such systems in extreme relative motion.

As a young adult, Einstein was attracted toward professional physicists, chemists, and mathematicians as models, although he never lost his youthful interest in the theory of knowledge. Soon methodological problems in physics led him into epistemological problems and beyond into ethics and aesthetics. More contemplative than his peers and more intuitive than his teachers, he was determined to publish his thoughts and to make a professional name for himself. As a struggling and idealistic young graduate student, he had to compromise his desires to remain a world citizen by becoming a Swiss national in order to find employment. He enjoyed the leisure that remained after fulfilling the duties of his patent-office job. It provided an interesting living and a vantage point for observing the swift march of technology without too much social involvement. With a growing family to care for, he could not afford to travel to professional meetings and conferences, and so he cultivated a circle of close friends where intimate communication could occur. His handful of intellectual companions plus a few fellow chess players and chamber musicians made for a stimulating though simple lifestyle. One real joy beginning in 1901 came from the occasional acceptance of his papers for publication in the foremost German journal of physics, *Annalen der Physik.*

As we shall see, the first, or restricted, theory of relativity was in the air in 1904 and yet, truly, it was also a personal creation. It sprang from a fertile mind more attuned to the central needs of advancing physics than to the broadcast fashions of the profession or the fads of the larger scientific community. After 1910 or so, as he became professionally recognized for his four percipient papers of 1905 and for more insights thereafter, Einstein became an authority, a judge within Germany's elite "invisible college" of theoretical physics. After the completion of his general theory of relativity in 1916, and after the nationalistic hatreds of 1914–18 subsided somewhat, Einstein, the Swiss pacifist amid German militants, became a culture hero for most of his profession and a celebrity outside of science as well as beyond Europe. He was so lionized in the 1920s that professional jealousy, often overcast by nationalism and underpinned by anti-Semitism, became a severe problem. Public adulation constantly threatened to disrupt his contemplative life. By travel before, and emigration after, the Nazi coup, Einstein managed to retain his simplicity, generosity, and balance in spite of the growing popular craze for relativity. Only Freudianism seemed as intriguing an intellectual fashion during the 1920s and 1930s.

Meanwhile physics developed apace, and Einstein remained committed to his search for a still higher synthesis, for a unified field theory. As the profession passed through the quantum mechanical revolution of the mid-1920s, it also passed well beyond Einstein's personal philosophical preferences. The master lived to see most of his contemporary disciples move toward the statistical "apostasy" of quantum mechanics and probabilism. Despite Einstein's deterministic beliefs, most physicists still revered him, though seldom extending themselves so far as to endorse his philosophical prejudices for continua or for a socialistic society. He, like everyone else, was caught in the maelstrom of world history, of war and peace, and of public versus private duties. When he died in 1955 after a half century of professional contributions, a generation of public pronouncements, and a decade of remorse for the nuclear weapons that had evolved from his mass-energy equivalence principle, he was eulogized as much for his active efforts in behalf of peace and world government as for his contributions to natural philosophy. The civilized world recognized one synonymy between physics and Einstein and another between pride in science and humility before the threat of nuclear apocalypse. The intellectual world knew it had lost a great man; just how unique and how much a part of his time he was, the world is yet to learn.[2]

In July 1952, Einstein wrote a foreword to a new translation of Galileo's Dialogue Concerning the Two Chief World Systems—Ptolemaic and Copernican, which reads in part as follows:

The theory of the immovable earth was based on the hypothesis that an abstract center of the universe exists. Supposedly, this center causes the fall of heavy bodies at the earth's surface, since material bodies have the tendency to approach the center of the universe as far as the earth's impenetrability permits. This leads to the approximately spherical shape of the earth.

Galileo opposes the introduction of this "nothing" (center of the universe) that is yet supposed to act on material bodies; he considers this quite unsatisfactory. . . .

Therefore, the hypothesis of the "center of the universe" had to be replaced by one which would explain the spherical shape of the stars and not only that of the earth. Galileo says quite clearly that there must exist some kind of interaction (tendency to mutual approach) of the matter constituting a star. The same cause has to be responsible (after relinquishing the "center of the universe") for the free fall of heavy bodies at the earth's surface.

Let me interpolate here that a close analogy exists between Galileo's rejection of the hypothesis of a center of the universe for the explanation of the fall of heavy bodies, and the rejection of the hypothesis of an inertial system for the explanation of the inertial behavior of matter. (The latter is the basis of the theory of general relativity.) Common to both hypotheses is the introduction of a conceptual object with the following properties:

(1). It is not assumed to be real, like ponderable matter (or a "field").
(2). It determines the behavior of real objects, but is in no way affected by them.

The introduction of such conceptual elements, though not exactly inadmissible from a purely logical point of view, is repugnant to the scientific instinct. . . .

Once the conception of the center of the universe had, with good reason, been rejected, the idea of the immovable earth, and generally, of an exceptional role of the earth, was deprived of its justification. The question of what, in describing the motion of heavenly bodies, should be considered "at rest" became thus a question of convenience.[3]

Einstein's recognition of the analogies between Galilean relativity and his own general theory of relativity may help us to recognize how closely allied were the minds of those two thinkers and their attitudes toward physical science and natural philosophy.

NOTES

1. Albert Einstein, "Autobiographical Notes," in Paul A. Schilpp, ed., *Albert Einstein: Philosopher-Scientist* (New York: Harper Torchbooks, 1959), pp. 3, 31, 53.
2. See, e.g., Bertrand Russell's Foreword to *Einstein on Peace*, Otto Nathan and Heinz Norden, eds. (London: Methuen, 1963). See also Notes for chaps. 4 and 5 of this book for further references.
3. Albert Einstein, Foreword to Galileo Galilei, *Dialogue Concerning the Two Chief World Systems—Ptolemaic and Copernican*, ed. and trans. Stillman Drake (Berkeley: Univ. of Calif. Press, 1967 [copyright 1953] from first edition, *Dialogo* . . . , Florence, 1632), pp. xi, xiii, xv; Foreword translated by Sonja Bargmann. Copyright 1953, 1962, and 1967 by The Regents of the University of California; quoted by permission of the University of California Press.

Chapter I

The Maxwellian
Synthesis
(c. 1870s)

> Our whole progress to this point may be described as a gradual
> development of the doctrine of relativity of all physical
> phenomena.
>
> —JAMES CLERK MAXWELL, 1876

Galileo Galilei died on 8 January 1642. Within a year Isaac Newton was born—on Christmas Day. On 14 March 1879 Albert Einstein was born. Later that same year, on 5 November, James Clerk Maxwell died. These four men above all others set the styles and provoked the studies that evolved into the "relativity" revolution of the twentieth century.

"In the beginning God created the heavens and the earth," begins one of the oldest authorities in the culture these four men shared. Long before that was written, wise men wondered what the myths of Genesis really meant. Each of these four shared that wonder. Galileo, Newton, Maxwell, and Einstein led the formation of a subculture or supraculture that redefined both the questions and the answers that religion and philosophy had traditionally tried to serve.

During the nineteenth century, however, the accumulation of

exact knowledge about what Earth is made of and how it moves had become so large that natural philosophy came to be called "scientific" knowledge, and its specialized practitioners became known as "scientists" instead of "men of science" or philosophers. The love of wisdom became so diluted by divisions into parts that few knowledgeable men dared to declare their love of wisdom as a whole.

In western universities during the nineteenth century, new disciplines were born and professional chairs or departments were created to accommodate such specialized knowledge. Students of the ultimate constitution of matter called themselves "chemists"; those who studied all forms of the motions of matter came to be called "physicists"; and students of life became "biologists." Soon even these categories proved much too large, and with each new generation scientists specialized more and more until few remained to espouse synoptic views. Despite all this division of labor, academic scientists were a proud breed of men who by the end of the nineteenth century generally thought they had arrived close to an ultimate understanding of what matter really is, what explains the relative motions of our world, and what links all life forms together. The artificial features of the environment in which they lived, including their own academic departments, were radically different from any previous human experience, as a result of several ongoing industrial and organizational revolutions. Those affected could hardly imagine any other cause for the progress of the nineteenth century except the reinforcements built up over time by the interaction of science and technology.[1]

In large measure these men seem correct in their assessment of their historical condition. But with the added advantage of another hundred years of hindsight, it is possible to discern a certain myopia in their viewpoint—and a certain hyperopia as well. They were shortsighted in their pride of achievement and in their belief that nature is so easily knowable. And they were too farsighted in their assessment of the short-range changes that their changed style of life and thought would force on their generation. This is to say that they were usually surprised by the slowness of social change and by the failure of the majority of mankind to adopt their "scientific" attitudes.

Not only scientists but historians and social critics also suffered from these defects of vision, one result of which was a furious intellectual debate over determinism versus voluntarism in the molding of historical change. Scientists themselves tended to see their intellectual labors as built freely upon an accumulation of individual contributions; whereas engineers and technologists were more likely to emphasize the determining role of social forces, economic pressures, and the statistical behavior of large groups of people. Neither the "great man" of Thomas Carlyle's theory of history nor the depersonalized social generalizations of Karl Marx and other determinists offered satisfying explanations to the majority of enlightened men, who were all infected by the nineteenth-century cult of progress.

London was the largest city in the western world at midcentury. It was the capital of the British Empire and thus also the economic and political center of the first truly global and conscious network of human communities. London advertised itself in the great exposition of 1851, the first of the "world's fairs," as the center of culture, commerce, and cosmopolitan hopes. In the 1860s, with a population of 3 million to 4 million, it harbored, occasionally at least, men as diverse as Charles Darwin, Herbert Spencer, Karl Marx, John Stuart Mill, and James Clerk Maxwell.

Maxwell was a rather modest mathematical physicist who was also a cultured Scottish gentleman. His lack of flamboyance and his regrettably short lifetime combined to deny him some of the fame that may have been his due. But despite his relative lack of public recognition even now, Maxwell deserves to be known as one of the greatest physicists of all time because of his insight into the unity of radiant heat, light, electricity, and magnetism. The Maxwellian synthesis gave birth to the electromagnetic spectrum. And that spectrum of radiation lies at the core of modern science. Einstein learned to know and honor Maxwell as his primary intellectual forebear. There were others as well—particularly Mach, Helmholtz, Hertz, and Lorentz—but only Maxwell and Lorentz were primarily physicists' physicists who sought to understand through unification of theories how electromagnetic nature behaves.

Modern studies of scientific revolutions have been profoundly influenced by debates within the philosophy of science over the foundations of scientific methods. Philosophical problems of induction and deduction and their corollaries as applied to all the different disciplines we call "scientific" are so disparate as to leave much doubt about the process of scientific change. But there seems to be a growing consensus that "normal science" proceeds primarily in puzzle-solving activities, whereas "revolutionary science" interrupts occasionally to resurrect perennial problems and redefine them in terms of new models for how best to "do science." This general model for the history of science assumes that normal analyses carried on by rank-and-file scientists constitute the core of communal achievements, whereas the greatest contributions are abnormal syntheses. The most significant discoveries and inventions are made, it seems, by individual minds in intuitive leaps of "genius" that provide new conjunctions of what were formerly considered separate phenomena. In retrospect Aristotle, Euclid, Archimedes, Ptolemy, Copernicus, Kepler, Galileo, Newton, Maxwell, and Einstein are most often honored as the greatest codifiers and synthesizers of the western tradition of physical science. Depending upon the size of the historical tapestry and the focal points chosen, the list of great men may vary and be expanded or contracted to fit the artist's frame. But no account of the history of modern physics can afford to ignore the roles of Galileo (1564–1642), Newton (1642–1727), and Maxwell (1831–79) in thinking through relative motion before Einstein.[2]

Galilean Relativity

Short of recalling all of Galileo's *Dialogo* (Florence, 1632), we may begin by paraphrasing almost in caricature the way Galileo himself might have analyzed the modern word "relativity." To be relative, two or more things in this universe must by definition be related. To be absolute is to be alone. Motion implies rest and some kind of relationship, some sort of interrelatedness, between things. Any kind of system isolated by constant movement with respect to its universal surroundings exhibits a mechanical relative motion with respect to those surroundings. Matter in motion, the old definition of physics, implies therefore mechanical relativity. And this relativity, by itself, precludes our ability—or Aristotle's or Einstein's—to ascertain precisely what point is the center of the universe. Wearing the double mask of Salviati and of Sagredo, Galileo devoted most of his *Dialogue Concerning the Two Chief World Systems— Ptolemaic and Copernican* to a discussion of the subtleties involved in celestial and terrestrial relative motions. What is true? What only apparent?

Thus when Galileo began to specify the law of falling bodies, illustrated in part by cannonballs fired from shore to ship or ship to shore and following parabolic trajectories, his analysis of such interrelationships became in fact the first "theory" of relativity. In effect Galileo taught that the laws of motion hold good for any "isolated" system of matter, if that system is merely moving uniformly with respect to its neighboring systems.

Galilean relativity, therefore, was a central insight of the scientific revolution. It marks a distinct departure from Aristotelian physics, from the impetus theorists, and from traditional western intuition. No doubt even Galileo's imagination could not have been sparked to consider mechanical relativity seriously without the stimuli provided by the oceanic galleons, ballistics-minded gunners, the mining industry, and the general intellectual ferment that came with the sixteenth-century age of exploration. Indeed he makes Sagredo lament, "What a shame there were no cannons in Aristotle's time!" Galileo's famous thought-experiments included one of dropping objects from the masts of a moving ship and noting the different ways in which objects appear to fall when viewed from the deck and from the dock.

Each variation of these and other experiments he discusses is designed to draw a tighter noose around the problems of relative motion and of appearances versus realities. Perhaps his best short expression of "relativity" is as follows: "Motion, in so far as it is and acts as motion, to that extent exists relatively to things that lack it; and among things which all share equally in any motion, it does not act, and is as if it did not exist."[3] Galileo agreed with the Aristotelian saying that "Whatever moves, moves with respect to something motionless," but he emphatically challenged the Aristotelian idea that "something motionless" is equivalent to "something immovable" upon which everything that is moved must be moved. "Motionlessness" may be a condition of relative rest with respect to something in motion, whereas "immovability" is an assumption of absolute position that is beyond human ken, or at least beyond physical or astronomical capabilities. "Metaphysics" may have its space full of absolutes, but the primary work of plain physics and astronomy is to observe what happens, test causes by experiment, save the appearances of relative motion with a minimum of coherent descriptions and corresponding explanations.

Galileo's synthesis has long been considered by many physical scientists as the greatest of all time. Although the heroic and dramatic elements in Galileo's life, trials, and influence have contributed much to the veneration in which he is held, the birth of the science of motion does owe him more than any other. Mechanics, rational and experimental studies of matter in motion, required the exposure of many inherited fallacies of a most fundamental sort. Galileo demonstrated his happy method of applying geometrical analysis to physical problems, then persuaded most of his readers with his felicitous literary style. He taught that gravity and levity are relative terms, that all bodies and fluids are heavy, that motion is the result of force, that weight is a continuous force, that inertia implies continuance of motion as well as permanence of rest, and that heavenly bodies must be as imperfect as earthly ones. All this and more he set forth in his *Discourses . . . Concerning Two New Sciences* (Leyden, 1638), half of which was devoted to strength of materials and half to local and compound motions of projectiles. Despite Galileo's neglect of Kepler, his mistakes, arrogance, and human failings, the Galilean synthesis deserves to be remembered as one of the grandest achievements in all history.[4]

Newtonian Relativity

Newton's *Philosophiae Naturalis Principia Mathematica* (London, 1687) began with a series of eight definitions of matter, motion, forces, and acceleration followed by a famous "Scholium," which distinguished between ideal or absolute *time, space, place,* and *motion* and the vulgar or commonsensical meanings of these terms. Having divided these four concepts into "absolute and relative, true and apparent, mathematical and common" meanings, Newton went on to wrestle for several famous pages with the most profound problems underlying his assumptions. Before stating his axioms, or three laws of motion, he declared that "in philosophical disquisitions, we ought to abstract from our senses, and consider things themselves, distinct from what are only sensible measures of them. For it may be that there is no body really at rest, to which the places and motions of others may be referred." After admitting the great difficulties involved in such speculations and in experimental attempts to make such distinctions, Newton finally said by way of introduction: "How we are to obtain the true motions from their causes, effects, and apparent differences, and the converse, shall be explained more at large in the following treatise. For to this end it was that I composed it."[5]

The principle of inertia and the idea of different, separate inertial systems, each of which may appear isolated when viewed from within and relative when compared with another, were central insights of the Galilean relativity that Newton recognized. Newtonian relativity, distinctly mechanical and dynamically difficult to measure with the techniques of the eighteenth century, was based directly on the explicit assumptions that absolute space and absolute time must exist in nature, even if only apprehensible to the mind of nature's creator, God. By moving on into theoretically derived explanations, mathematically certified and empirically tested, Newton achieved immortality by his synthesis of the laws of celestial and terrestrial motion. Kepler and Galileo had set the planetary firmament and experiential gravity in order, respectively, so that Newton could stand on their shoulders to convene his laws of motion and a theory of universal gravitation.

If Newton hid his "fluxions" or differential calculus when composing his *Principia* in classical geometrical style, he flaunted his four "Rules of Reasoning." He framed a fifth rule regarding hypotheses but later decided to suppress it. If he fudged on his figures to clinch his case, he also molded his laws of inertia, force, and action to fit the dynamics of point-forces, inverse-squares, and attraction-at-a-distance. He often changed his mind, as the variorum edition now shows, but the severely classical and Latinate *Principia* was finally committed to the future with the impression that others would come to perfect its imperfections and its "System of the World." One of these failings concerned the *cause* of gravity, about which Newton tried to feign no hypotheses. The ancient quintessence of a most perfect, subtle, and ubiquitous aether had served for some, like Descartes and Huygens, this purpose of filling up space, transmitting light through emptiness, and perhaps, causing all bodies to be attracted to one another by a force inversely proportional to the square of their distances apart. But Newton left to his popular book of chemistry and queries, *Opticks* in vernacular English, the speculations that the aether as *Sensorium Dei* might also serve many other functions besides bearing light through empty space from sun, moon, and the stars to lenses, prisms, and eyes.[6]

Newton's own ambiguity over the role of continuous versus discrete modes of explanation for the behavior of light was largely forgotten by his disciples. Newtonians grew ever more confident, even if quite divergent in different areas and disciplines, as their mechanical world-view and corpuscularian philosophy generally seemed well corroborated by the events of the eighteenth century. If the Enlightenment and Age of Reason worshiped Newton uncritically, perhaps most Newtonians did so because their critical eyes were cocked elsewhere. In any case, relative motion reappeared in many different guises.[7]

James Bradley (1693–1762) was an English divine who became professor of astronomy at Oxford in 1721 and soon thereafter became interested in the problem of stellar parallax. Together with Samuel Molyneaux, he set up a rather elaborate zenith telescope to test for parallax (apparent shifts in positions of stars due only to Earth's orbital motion) at four seasons of the year. But surprisingly, Bradley found both less and more than he bargained for, his target stars behaving contrary to expectations. After several years of observations, thought, and calculations, Bradley announced his discovery of astronomical aberration in 1729, which he explained as a result of the finite speed of light and of Earth's motion in orbit. From residuals remaining, after two more decades Bradley also discovered the nutation of Earth's axis and the "variation of latitude," but still no stellar parallax. He was named the third Astronomer Royal in 1742. Though a good Newtonian, Bradley left Earth more unstable than he found it.[8]

Four other Newtonian astronomers of the eighteenth century helped to expand the scale, though not to challenge the basis, of Newton's assumptions about the universe.

Thomas Wright (1711–86) of Durham was first to surmise the general disklike structure of our galaxy by analysis of the Milky Way as a relatively finite creation though "formed of an infinite number of small Stars." Immanuel Kant (1724–1804) shortly thereafter used Wright's "Original Theory . . . of the Universe" (1750) to propose his own "Theory of the Heavens" (1755). Thinking of the system of the Milky Way as being replicated many times in "nebulous" stars, Kant was moved to propose the existence, beyond the innumerable systems of this galaxy, of an infinite multitude of other systems and other galaxies. Later Kant's cosmology became known as the "island universe" hypothesis, but his redefinition of the center-of-the-universe problem was less acute than that of a contemporary self-taught mathematician.

Johann Heinrich Lambert (1728–77) working independently on a hierarchical system of celestial systems was profoundly impressed by the problem of universal motion, relative and absolute, as might be seen from the center of centers, "the center of creation, which I should be inclined to term the capital of the universe." In his *Cosmological Letters*, he wrote:

But who would be competent to measure the space and time which all the globes, all the worlds, all the worlds of worlds employ in revolving around that immense body, the Throne of Nature, and the Footstool of Divinity? What painter, what poet, what imagination is sufficiently exalted to describe the beauty, the magnificence, the grandeur of this source of all that is beautiful, great, magnificent, and from which order and harmony flow in eternal streams through the whole bounds of the universe?[9]

Lambert asked where might be the center of the Milky Way itself in relation to all the other milky ways, and he insisted that Kepler's ellipses must be seen as cycloid trajectories on ever larger frames of reference:

The Moon, it is said, describes an ellipse round the Earth. This would be true were the Earth at rest; but as she moves round the Sun, and obliges the Moon to participate in her motion, the orbit of this last cannot be an ellipse, but a cycloid. The ellipse of the Earth vanishes for the same reason, the moment the Sun ceases to be immoveable, and is found to describe an orbit round a new centre. Then the ellipse of the Earth becomes a cycloid of the first degree, that of the Moon of the second, and the velocity of their motion increases in the same proportion.

Carrying on this analysis of epicycloids indefinitely, Lambert came to rest with this passage about absolute motion, still within the Newtonian assumptions:

As we pass on from centre to centre, these motions become more and more complicated; and their combinations only terminate at the universal centre, which alone is in a state of real and absolute rest. If, beginning by the Moon, we suppose that the body which occupies that centre is in the thousandth; the cycloid of the Earth will be in the nine hundred and ninety-eighth [sic, ninety-ninth] degree. There, and there alone, will be the true orbit of the Earth, while the velocity with which she describes it, will be her true velocity.

Lambert despaired of finding anyone who could make such determinations, and although the probability of ever knowing the final truth about Earth's ultimate cycloid was slight, Lambert believed in progress, and perhaps "twenty centuries hence" a much better approximation to that truth might be in hand. Lambert was unaware that another German gone to England would take these ideas as one immediate challenge for observational astronomy.

Frederick William Herschel (1738–1822), a musician from Hannover who emigrated to England, an amateur optician, and soon a professional astronomer without equal, began his telescope-making and heavenly observations during the 1770s, the decade of the American Revolution. In 1781, he discovered the planet Uranus and its satellites, and by 1782 he had started his masterful catalog of the heavens in the Northern Hemisphere, mapping double stars especially, with a view toward the determination of the relative distances of the stars from our sun and from one another. George III then appointed Herschel to be his private astronomer, and his remuneration for this allowed him freedom to pursue his second career more systematically and methodically than ever before.

In 1783, Herschel produced his sublime speculation on the "Motion of the Solar System in Space," a theme to which he returned in 1805 after getting well into the problem of charting the Milky Way and mapping our sun's relative position within its galaxy. When finally knighted in 1816, Sir William had convinced himself, with the aid of his giant reflector (of 40-foot focal length and 4-foot aperture) as well as other telescopes, that the Milky Way was essentially "fathomless." And yet that our solar system has intrinsic motion against the so-called fixed stars of the celestial sphere was seldom doubted thereafter. Herschel finally gave up determining the sun's absolute speed, but his direction for the solar apex (or azimuth for the sun's translation) was long accepted as valid. Herschel felt the solar system had come from the neighborhood of Sirius and was headed toward Hercules at perhaps 5 km per second. The evidence for the direction of such an "absolute motion" came from the star catalogs on positional perspective, which seemed to indicate gradual closure of "fixed stars" in the neighborhood of Sirius, and in the antipodal direction of the neighborhood of Hercules a gradual widening of the positions of "fixed stars."[10]

Sir William's son, John F. W. Herschel (1792–1871), finished his father's work with a star catalog for the Southern Hemisphere and with many more contributions to mathematics, physics, and astronomy. Yet to Sir William goes the credit for discovering, in 1800, the infrared region of the optical spectrum. The next year similar radiations were found beyond the violet end of the spectrum, and these extensions of Newton's theory of optics set the stage for the work of Thomas Young (1773–1829) in championing the wave theory of light. At the beginning of the nineteenth century this seemed a most non-Newtonian thing to do.

Electricity and Magnetism—Ancient Enigmas

Since the dawn of consciousness, lightning and lodestones have fascinated and terrified people. Yet so different are sky-sparks and iron magnets that it is hardly surprising it took so long to surmise some connection between the two. Lodestones as magnets were named by the Romans, and the Greeks had prized amber *(elektron)*, a petrified resin, for its attractive yellow color and its property of attracting particles of dust and lint after being vigorously rubbed. But few are the records of further interest in such mysteries before 1600. William Gilbert (1544–1603) was appointed physician to Queen Elizabeth in 1600 and also that year published his treatise *De Magnete,* one of the greatest classics of experimental science. Gilbert studied both electrical and magnetic phenomena, suspecting some profound relationship; but he was content merely to show how magnetism can be increased, decreased or destroyed, and transferred from lodestones to iron bars. That the whole earth behaves like a giant magnet and that many other substances besides amber can be electrified by friction also were his discoveries. Galileo was impressed, as was Kepler and many others. But not until the eighteenth century did experiments with static electricity really catch hold.

Benjamin Franklin (1706–90) took another giant step with his kite experiment of 1752, after seven years of studying electrical charges and discharges. He proved that lightning is electrical in nature by conducting atmospheric energy down a wet string to charge a Leyden jar, produce sparks, and duplicate the other curiosities of static charges. His "single fluid theory," using + or − descriptions of state, eventually replaced the more popular two-fluid theory of electricity and thus helped pave the way for understanding dynamic electricity.

Charles Augustin de Coulomb (1736–1806) before and after the French Revolution was perhaps the foremost scientist of static electricity and magnetism. His fundamental memoirs established many quantitative relationships, and his torsion balance became the prime instrument to measure electric and magnetic forces of repulsion and attraction. By 1801, Coulomb had established in full generality his famous law of force for both electrostatics and magnetism. It closely resembled Newton's law of gravity, but Coulomb preferred to retain a two-fluid explanation.[11]

The man most often credited with inaugurating the study of electrodynamics—electricity in motion—is Luigi Galvani (1737–98), anatomy professor at Bologna and discoverer of "animal electricity." Observing frogs' legs twitching under various dissection arrangements in 1786, Galvani was fascinated and experimented broadly to convince himself by 1791 that frogs' legs were like Leyden jars when the frogs themselves (or their parts) were part of a circuit containing pieces of metal. Alessandro Volta (1745–1827), professor of physics at Pavia, vigorously disputed Galvani's claims and plunged immediately into trying to explain what had really happened in Galvani's experiments. Convinced that the animals merely provided moisture, not electricity itself, for completing an electrical circuit, Volta experimented for almost a decade with different pairs of metals, trying to produce "metallic electricity" without animal matter. Finally, in 1799, he succeeded in creating his "voltaic pile," an electrochemical series of silver and zinc disks separated by pasteboard insulators soaked in a salt solution. This, the world's first electrical battery, was announced in a letter in 1800 to the president of the Royal Society, London. It was to revolutionize the study of electricity, because here was the world's first steady supply of current electricity. Electrolysis began almost immediately, as water was decomposed by use of Volta's batteries and as electroplating studies followed thereafter. Galvanic currents and voltaic cells soon became quite fashionable, and electrochemical discoveries tumbled over each other, as Humphry Davy (1778–1829) created the "voltaic arc" light.[12]

At the beginning of the nineteenth century, thanks largely to the discovery of interference as a property of light by Thomas Young and its mathematical elucidation by Augustin Fresnel (1788–1827), the corpuscularian theory of light began to be superseded by the undulatory theory. Lagrange (1736–1813) and Laplace (1749–1827), meanwhile, were putting capstones on Newtonian theory in analytical and celestial mechanics; Lavoisier, Priestley, Cavendish, and Dalton had been and were ushering in a revolution in chemistry; and Coulomb, Galvani, Volta, and Davy were among those converting studies in electricity from static to dynamic modes of physical action.

Leonhard Euler (1707–83), the most prolific of all mathematicians, was such a prodigy that the eighteenth century can hardly be mentioned without paying him homage. He was an algebraist, geometrician, and analyst whose collected works in their most recent edition will comprise seventy-four volumes. His fluid mechanics had to await the next century to be properly appreciated, for he was the only major theorist of the eighteenth century to favor the wave theory of light and thus the luminiferous aether. Euler lived much of his life in Prussia and Russia, far removed from the mainstream of European physical theory. Yet his methods and examples provided many problems and proofs that fed into the heritage of Gauss and Cauchy.[13]

Wave Theory and Electromagnetism

Augustin Louis Cauchy (1789–1857) was one of many young Frenchmen who benefited directly from the Napoleonic reforms in education. Trained as an engineer, he was soon persuaded by family friends Lagrange and Laplace to devote himself entirely to mathematics. This he did with marvelous success, over wide ranges of mathematics, astronomy, and physics, despite the many vicissitudes of France's political history during his time. After several elegant solutions to some ancient problems in his youth, Cauchy first attracted wide notice with his study on wave propagation, which won the 1816 grand prize of the Academy. His treatises on analysis, the calculus, and their applications to geometry established new methods of great rigor. Perspicuous and prolific, Cauchy published 789 papers, which fill twenty-six volumes in his *Oeuvres Complètes* (1882–1938).

Karl Friedrich Gauss (1777–1855) is generally considered the greatest mathematician of all time. Both Cauchy and Gauss were interested in the ideas of convergent series, and both contributed heavily to applied mathematics. If Gauss had no equal in the theory of numbers and perhaps none in curvilinear coordinates and conformal mapping, his stature as a versatile and universal genius was discovered only after his death, from unpublished papers. During the lifetimes of these two, Cauchy seemed to be the more universal man. In mechanics, Cauchy substituted the notion of continuity of geometrical displacements for the principle of the continuity of matter. In optics, he developed the wave theory and a dispersion formula for undulatory propagations. In elasticity, he originated the theory of stress. But, unlike S. D. Poisson (1781–1840), whose influential contributions on electricity and magnetism were based on mathematics analogous to the theory of gravitational attraction, Cauchy contributed directly to the growing mathematical support for the undulatory theory of light.

Gauss, the prince of mathematical rigor, might have been recognized as a monarch had he not been so reticent and true to his motto, *Pauca sed matura* (few, but ripe). A child prodigy who never lost his prodigious talents, Gauss was truly a polymath who contributed basic insights to every field of mathematics. Looking askance at the bases of Euclidean geometry since childhood, he fostered by word of mouth and mind the birth of non-Euclidean geometries. His classic treatise *Theoria motus corporum coelestium* (1809) codified the methods of perturbation theory for astronomy. Wilhelm E. Weber (1804–91) was one of three brothers who lived near Gauss in Göttingen and with whom he collaborated in studying the mathematical basis for electrical and magnetic phenomena. Gauss's encouragement and praise for the work of G. F. Bernhard Riemann (1826–66) led both to be posthumously honored for their extensions of the differential geometries of curved surfaces and spaces of n-dimensions. Perhaps Riemann's early death prevented their being recognized as mathematical physicists dealing with electromagnetism in its most general sense.[14]

The uses of many instruments and interferometric methods to study the radiant energy of the visible spectrum were readily adaptable from acoustics to optics. These two disciplines, trying to make scientific sense out of sound and sight, had expanded their boundaries beyond the range of direct human perception in the early years of the century. Interferometric techniques were based on the presumption of wave motion in a continuous medium, and this in turn assumed a periodicity of displacements in some medium. Waves could be superimposed to reinforce or cancel each other depending upon their phase at a particular point. The acoustic analogy had grown strong early in the nineteenth century when Thomas Young reinvigorated the wave theory of light with his experimental demonstrations of interference, showing how light added to light can produce darkness. But shortly thereafter, studies of polarization phenomena led to the firm assurance that light is transmitted by *transverse* waves, rather than by longitudinal waves as in the case of sound. Etienne-Louis Malus (1775–1812) and Jean Baptiste Biot (1774–1862) each contributed much to the birth of polarimetry, but D. F. J. Arago (1786–1853) was perhaps the most influential astronomer-physicist because his work in electromagnetism and light laid the bases for crucial tests of the wave theory later performed by others.

In the 1820s, after J. B. J. Fourier (1768–1830) and A. J. Fresnel developed comprehensive mathematical models for heat and light, respectively, wave theories grew ever more respectable and powerful. By midcentury, the wave theory of light had successfully accounted for reflection, refraction, diffraction, interference, and polarization. Then the arch rivals A. H. L. Fizeau (1819–96) and Leon Foucault (1819–68) at last succeeded in making two different terrestrial measurements of the speed of light, proving it to be manipulable and finite. Foucault shortly thereafter administered a coup de grace to the corpuscular theory by performing a supposedly "crucial" experiment showing that light travels more slowly in water than in air. According to the Newtonian emission theory, light should actually travel faster in a denser medium. Fizeau followed up later in the 1850s with a test of the velocity of light in flowing water, showing that Fresnel's theory was correct in predicting a partial drag: light moved faster with the current than against it. Thus the acoustic analogy came to be recognized as merely a crude approximation for the behavior of light. And although the transverse wave theory seemed to require an elastic-solid medium filling all space and transparent materials, the experimental evidence for a wave theory of light seemed overwhelming by 1860 and remained so through most of the 1880s.

Despite Maxwell, the optical analogy for the behavior of electromagnetic waves was widely regarded until the mid-1890s as possibly equally misleading. Although Kelvin, Faraday, Maxwell himself, and other physicists often tried to develop mechanical analogies based on fluid mechanics and hydrodynamics, such efforts were never quite satisfactory, primarily because of the lack of a link between aether and matter. The luminiferous aether as a kind of universal substratum was usually regarded cautiously as a temporary conceptual necessity whose physical properties would someday be delineated.[15]

Fresnel's work on the wave theory of light also led into his wave theory of heat, which enjoyed quite a vogue for several decades. But studies that attempted to synthesize light and heat, sound and stress, or light and sound were not very successful, because new anomalies from experimental experience kept arising to show that nature is more complicated than compromising. Even the seemingly simple wave theory of light and sound had to be segmented because sound waves are longitudinal compressions and rarefactions of their medium whereas light waves were proved to be transverse vibrations once polarization phenomena were thoroughly analyzed. Thus, toward the end of Cauchy's and Fresnel's lives the medium required for light seemed to be an elastic solid, very tenuous and subtle to be sure, but still a *solid* rather than a gas or a liquid. Many mathematical physicists grew old in despair over this conundrum.

Hans Christian Oersted (1777–1851), perhaps more than any other person, bridged the gaps that existed in his time between electrical and magnetic phenomena and between experimental-mechanical and rational-idealistic philosophies of science, laying the foundations for the origins of electromagnetic field theory. As a young Dane interested in the new chemistry and in the *Naturphilosophie* growing out of the influence of Kant and Schelling, Oersted tried to reconcile the conflicting forces of attraction and repulsion well before distinctions between kinetic and potential energy became common. In a classroom physics demonstration he stumbled almost accidentally upon the connection between an electric current and magnetic force. In July 1820, Oersted published the results of his now famous experiment that showed the deflection of a compass needle by bringing it near a wire carrying a current. This discovery of a direct connection between electricity and magnetism reinforced a belief he had held at least since 1812, allowing him to state unequivocally what he called the "fundamental law of electromagnetism": "that the magnetical effect of the electric current has a circular motion around it." By setting these complementary forces in the space surrounding their material bodies, Oersted gave birth to the field concept.

George Simon Ohm (1787–1854) was head of mathematics and physics at the Polytechnic in Cologne when in 1827 he announced the empirical law of resistance that came to bear his name. Having tested various conductors and insulators for years trying to isolate their limits and catalog their properties, Ohm was well prepared to recognize all the factors that impede the flow of an electric current. He consciously used the suggestive hydrodynamic analogy for his electrodynamics, and he extended his results originally taken from thermoelectric couples to voltaic cells. Electromotive force (E) is equal to the product of the intensity of current flowing (I) multiplied by the sum of the resistance (R) offered by the whole circuit: therefore,

$$I = \frac{E}{R}$$

or as taught today, "amps" equals "volts" over "ohms." But Ohm's law was slow to be recognized for its value as a major milestone in the codification of the rules for electrodynamic behavior, partially because its empirical status seemed to have no necessary connection with prevalent electromagnetic doctrines.[16]

By 1830, immense interest was aroused in the implications of Oersted's electromagnetism. André M. Ampère (1775–1836) began quickly to focus his attention on these and other electrical phenomena, eventually deriving a formula for the mutual interaction between any two small elements of a conductor in terms of the magnitudes, separation, and relative orientation of the currents. Ampère's mathematical deductions from his own and other experimental facts followed consciously the Newtonian model for the formulation of the inverse square law of gravitation.[17]

Meanwhile, Joseph Henry (1797–1878) in Albany, New York, was building bigger and better electromagnets, and Michael Faraday

(1791–1867) in London at the Royal Institution was doing the same—and much more. Both independently discovered the phenomena of electromagnetic self-induction, thus paving the way for the invention of telegraphs, dynamos, electrical generators, and motors. But Faraday's strategic position, his prior publication in 1831, and his unquenchable curiosity led to his getting the main credit for discovering the implications of magnets and currents in relative motion. For three more decades, Faraday contested with Ampère and other electrical theorists for better explanations of electrical and magnetic phenomena. Faraday's laws of electrolysis (1834), specific inductive capacitance (1837), and rotation of plane-polarized light (1846?) led him ever deeper into the belief that real electromagnetic forces exist as certainly in the space surrounding as in place internal to, say, an iron bar magnet. This emphasis on three-dimensional "lines of force" permeating the neighborhood around magnets and moving currents was most heretical at the time, but it came to be the fundamental idea of classical field theory.[18]

William Thomson (1824–1907), who was knighted in 1866 and raised to the peerage in 1892 as Baron Kelvin of Largs, was an extremely precocious son of a Scottish-Irish professor of mathematics at Belfast and Glasgow. In 1845, he encouraged Faraday to concentrate on electrodynamic rather than electrostatic research. And this shift of concern led to some fruitful analogies with thermodynamics and the elastic-solid theory of the luminiferous aether. While Thomson (whom I shall hereafter call Kelvin) and Faraday cross-fertilized each other's imaginations, Faraday continued without mathematical aid or insight to develop his experimental understanding of electric and magnetic fields of force. Kelvin, seemingly everywhere at once, was using his mathematical expertise to develop an absolute scale of temperature, a theory of heat, a second law of thermodynamics, and a vector notation for expressing the energy of a system in terms of a volume integral throughout space. By finding the same results applied both to systems of permanent or temporary magnets and to systems of circuits carrying steady currents, Kelvin suggested that energy could be stored throughout these fields of force.

Kelvin played a central role in the midcentury clarification of the concepts of force and energy—dividing the "ability to do work" into kinetic and potential spheres—and in the development of dynamical theories for heat, light, fluids, solids, and gases. He was almost unbelievably prolific, and he became widely recognized as the acknowledged leader of British physics during his fifty-three years as professor of natural philosophy at Glasgow. But somehow he mistrusted the essence of Faraday's thrust. The wave theories of heat and light were part of the ambient atmosphere by midcentury, and Faraday in trying to understand the mysteries of electricity tended to emphasize both Boscovichian atoms (to dematerialize matter) and vibrating waves in some sort of medium that could carry his lines of force. Kelvin, caught up in the prevailing mechanism and materialism of his culture, sought always for dynamical models to explain kinematical facts. Kelvin tried hard and long to rationalize a luminiferous or electromagnetic aether; Faraday sought to expunge the idea of an aether as an imponderable fluid or elastic solid. Instead Faraday wished to fill all space with a three-dimensional web of lines of force. This desire required a new collaborator to make it acceptable, one who perhaps could take the best from Faraday and Kelvin for a new synthesis.[19]

Faraday, Maxwell, and Fields of Force

Just as one young Scot, later Lord Kelvin, had inspired Michael Faraday to rise from despair in 1845, so also another young Scot, James Clerk Maxwell, encouraged Faraday in a now famous lengthy letter to persevere with his accumulated researches. Maxwell wrote to Faraday on 9 November 1857, saying in part:

Now as far as I know you are the first person in whom the idea of bodies acting at a distance by throwing the surrounding medium into a state of constraint has arisen, as a principle to be actually believed in. . . . nothing is clearer than your descriptions of all sources of force keeping up a state of energy in all that surrounds them. . . . You seem to see the lines of force curving around obstacles and driving plumb at conductors and swerving towards certain directions in crystals, and carrying with them everywhere the same amount of attractive power spread wider or denser as the lines widen or contract.

You have also seen that the great mystery is not how like bodies repel and unlike attract but how like bodies attract (by gravitation). But if you can get over that difficulty . . . then your lines of force can "weave a web across the sky" and lead the stars in their courses without any necessarily immediate connection with the objects of their attraction.[20]

By the date of this letter, Maxwell had already begun his campaign to rescue Faraday's system and the field concept by translating them into respectable mathematical form. Beginning in December 1855, Maxwell had read a series of papers to the Cambridge Philosophical Society, "On Faraday's Lines of Force," which conciliated more orthodox physicists by converting Faraday's lines into "tubes of force" carrying an ideal fluid wherein the energy, potential, and work of the sytem resided, ready for calculation by more traditional methods. Soon thereafter Maxwell compared Ampère's system with Faraday's in a paper, "On Faraday's 'Electro-tonic State,' " which cautiously suggested the superiority of the English over the Continental approach, especially for trying to understand the surrounding medium. Significantly however, Maxwell's letter to Faraday in November 1857 expressed the chief wonder they shared about the nature of electromagnetism as compared with gravitation: the first of the great questions is "Does it require time?" Is gravity propagated or simply pervasive?

Then in 1861, as Faraday began a steady and irrevocable decline with age, Maxwell delivered his notable essay "On Physical Lines of Force," which contained an ingenious molecular-vortex model of the electromagnetic field and an interrelated yet preliminary theory of light. Several other papers throughout the decade, especially in 1863, 1865, and 1868, confirmed, expanded, and improved Maxwell's field equations and electrodynamical formulations. Thus his electromagnetic theory of light was essentially complete by the end of the 1860s.[21]

Heavily influenced also by Kelvin, Maxwell seemed to be growing almost as prolific, showing a most fertile mind and careful strategy in mapping out whole fields in the kinetic theory of gases, thermodynamics, fluid mechanics, color vision, and atomic-molecular theory. By general agreement now, Maxwell's greatest work was that done on electromagnetism and light in the 1860s, culminating in the publication in 1873 of the first edition of his famous *Treatise on Electricity and Magnetism*. Maxwell's grand achievement was to link the theories of electricity and of magnetism in a common bond with the undulatory theory of light. By giving quantifiable mathematical form to the empirical and qualitative insights of Michael Faraday, Maxwell was able to create a proper field theory. His differential equations were to give operational meaning to what Faraday had called the "electrotonic state" of spaces surrounding electrified and magnetic bodies. After Maxwell, electromagnetism became a single word, not even hyphenated, and electromagnetic energy transfer became calculable in terms of partial differential equations and classical Newtonian mechanics. By envisioning light as a small visible segment of a grand electromagnetic spectrum of radiation phenomena, Maxwell laid a new foundation and designed a new story for the mansion of physics.

It is sometimes said that of the three greatest British physicists of the nineteenth century—Faraday, Kelvin, and Maxwell—only the last was a physicist's physicist. Faraday and Kelvin both catered to public audiences, became internationally famous, and took on many consulting roles. Faraday improved lighthouses, among other services for the maritime British Empire, but he never sought riches (largely because of his ascetic religious beliefs), whereas Kelvin gained wealth and his title largely for his work on communications systems and the transatlantic cable. Maxwell, on the other hand, moderately wealthy by inheritance, lived an academic life closer to "pure" physics and anticipated developments at the vanguard of professional interests, which later proved to be central concerns. Nowhere is this more evident than in his main achievement in electrodynamics, the "electromagnetic theory of light." By subsuming optics under electromagnetism, he laid the foundations for a whole new realm for scientific exploration and technological exploitation. Whereas Faraday and Kelvin had helped the entrepreneurs of their age in the development of better transportation and communication systems, Maxwell, was more aristocratic and was interested in history, poetry, and fine arts, and able to foresee the architecture of future physical science.[22]

Maxwell's Life and His Major Rival

Maxwell was born to a distinguished Scottish family on 13 November 1831. By 1850, he had progressed through the academy and university in his home city of Edinburgh and had contributed several papers to scientific journals. He too had been a remarkably bright child in mechanics and mathematics, and his record as a student at Cambridge University was no less distinguished than Kelvin's. His playful yet profound interest in optics, colors, electricity, heat, and gas phenomena led him to experiment as well as to theorize in novel ways to learn the "particular go" of things.

While at Edinburgh, Maxwell was most influenced perhaps by Professor J. D. Forbes toward natural philosophy and by Professor William Hamilton toward a liberal Christian metaphysics. At the end of his undergraduate career at Cambridge, Maxwell began reading the whole corpus of Faraday's works. Soon after studying Kelvin's analogous treatment of electric force with heat flux, Maxwell began to publish the series of papers that were eventually to overturn the idea of action-at-a-distance in favor of interacting fields of force. But the achievement of a dynamic theory of electromagnetism required much more than merely adapting Faraday's idea of space permeated by lines of force to a generalized Lagrangian coordinate system described by partial differential equations.[23]

Maxwell served as professor of natural philosophy at Aberdeen, 1856–60, then as professor of physics and astronomy at King's College, London. There from 1860 to 1868, Maxwell did his best work in the calculus for electromagnetism. After a few years in retirement at his family estate, Glenlair, he was offered in 1871 the first chair in experimental physics at Cambridge. There he meticulously planned the development of the newly founded Cavendish Laboratory. He also edited the papers of Henry Cavendish for publication. In addition to his writings covering a wide range of scientific interests—from the stability of Saturn's rings to the theory of color vision—Maxwell produced a textbook on the *Theory of Heat* in 1871, published his fully developed *Treatise on Electricity and Magnetism* in 1873, and wrote an excellent introduction to science education, *Matter in Motion*, in 1876.

During his fertile years at King's College, London, Maxwell

Galileo Galilei *(courtesy The Smithsonian Institution)*

Isaac Newton *(courtesy The Smithsonian Institution)*

Michael Faraday *(courtesy The Smith-sonian Institution)*

James Clerk Maxwell *(courtesy AIP Niels Bohr Library)*

thought through all the analogies, experimental data, and logical formalisms that might apply to electrostatics and electromagnetism. The result was clear by 1867: he would take, cautiously to be sure, the basic idea underlying the undulatory theory of light, namely, its luminiferous aether, and allow that medium to serve as the seat for both potential and kinetic electrical and magnetic energy. Neither the name "aether" nor the notions of fluid flow should be invested with much trust, however. "Electromagnetic field" would better describe, if not explain, the neighborhood of space surrounding charged, or magnetized, bodies or "live" wires. The "energy of the field," literally to Faraday and Maxwell, became as real as the tangible bodies or wires at its core.[24]

Maxwell probably gained the first inspiration for his synthesis when he learned that Wilhelm E. Weber (1804–91) and R. H. A. Kohlrausch (1809–58) had determined the ratio of electrostatic and electromagnetic units-of-quantity by discharges of a condenser through a ballistic galvanometer. Because the ratio, announced in the late 1850s, was almost the same as the velocity of light, measured by Fizeau, Foucault, and others, a link between optics and electricity could be perceived. But to conceive a system of unification was quite a different problem. As early as 1833, Weber and Gauss had connected their respective laboratory and observatory at Göttingen with an electromagnetic telegraph. And Weber continued after Gauss's death in 1855 to amplify Ampère's approach to electrodynamics and Gauss's interests in magnetism without making this fundamental linkage. Weber thought of electricity in abstract terms as particles of opposite polarity that almost instantaneously switched places whenever currents flowed or sparks jumped across gaps in circuits. By the time Maxwell matured, Weber was the leading electrical theorist in Europe. Yet he and other German and French scientists were more impressed by the differences than by the similarities between optics and electrodynamics.

Weber's authority derived in part from his direct mathematical lineage within the Newtonian program, which used *ordinary* differential equations to express the action of electrodynamic forces acting *instantaneously* at a distance. As developed by Laplace, Poisson, Ampère, Gauss, and Weber, this approach virtually ignored the role of time in propagating electromagnetic effects. Because for most practical purposes the idea of instantaneity could be assumed without penalty, Weber's theory seemed fully adequate. But Maxwell's theory differed on just this point: as a mathematical synthesis of Faraday's intuitive dislike for instantaneous actions-at-a-distance, Maxwell used *partial* differential equations to express the behavior of a potential at every point in a field of space changing continuously through time. To Maxwell the first of the great questions about the mystery of gravitation—Does it require time?—related directly to the systematic affirmative answers he had provided in his theory of optics and electromagnetism. To Weber and most theorists on the Continent, however, it seemed more important to keep electrodynamics as simple as possible. Closed circuits and the atomic conception of electricity championed by Weber did not jibe well with Maxwell's displacement current and field equations. There seemed as yet no compelling need for a connection of the wave theory of light with electromagnetism.

But Maxwell following Faraday could and did conceive this linkage as a fundamental property of all electromagnetic field effects. By treating the electromagnetic potential as primary while making the mass and charge of a particle secondary, Maxwell was able to find "a mathematical method in which we proceed from the whole to the parts instead of from the parts to the whole." Ultimately this required him to mix his metaphors somewhat (the "vector potential" and the "displacement current" ideas were extremely difficult to sell to other experts), and to handle his mechanical and material models of the aether, for instance, as mere mental aids. Despite his many disavowals, however, non-English readers found Maxwell's models difficult to accept, and his mathematics therefore was often suspect.[25]

Maxwell published all together about one hundred papers and four books during his short lifetime. His work in the kinetic theory of gases and in thermodynamics was as significant and influential as that in color, sight, light, and electromagnetism. But Maxwell's reputation, gained posthumously as the main link between Faraday and Hertz, rests primarily on his five electrical papers of the 1860s and the subsequent *Treatise* of 1873. Although he never predicted that electromagnetic waves would be found to permeate space and behave exactly like radiant light and heat, that was the crowning glory that derived from his synthesis.

(The graphic representation of Maxwell's lifetime production record in Figures I and II (p. 34) shows how his publications were spread over his thirty years of activity. The purpose of the graphs is not well served by Niven's editing of Maxwell's *Scientific Papers*, but in order to compare the characters and reputations of great men in physics or science generally, it is helpful to contrast such data against others. Figures III and IV, in the Epilogue, allow such a comparison for the two principal men of this narrative.)

But little of this was recognized while he still lived. Maxwell died in 1879 at the age of forty-eight, almost a decade before the brilliant electric-spark experiments of Heinrich Hertz were to discover wireless waves and amplify the electromagnetic synthesis.

Maxwellian Relativity

In *Matter in Motion*, his guidebook to the study of physical science, Maxwell included eight chapters dealing with what he then regarded as the most fundamental ideas in physics ("matter," significantly, was not so regarded). The concepts of motion, force, work, energy, and gravity received the bulk of his attention, and a critique of the Newtonian tradition in physics was central to his exposition. Regarding Newton's explicit assumptions about absolute time and space, Maxwell stated simply, "All our knowledge, both of time and place, is essentially relative." He had learned this lesson from his mentor in metaphysics at Edinburgh, Sir William Hamilton (1788–1856), whom Maxwell revered, although he respected even more an unrelated Irish mathematician, Sir William Rowan Hamilton (1805–65). In recapitulating the analysis that followed that statement, Maxwell penned a section entitled "Relativity of Dynamical Knowledge." So cogent is this for our story that it is worth quoting at length:

Our whole progress up to this point may be described as a gradual development of the doctrine of relativity of all physical phenomena. Position we must evidently acknowledge to be relative, for we cannot describe the position of a body in

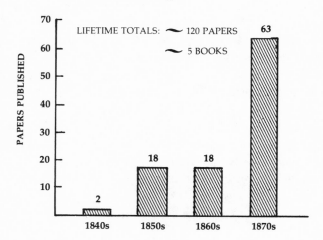

Figure I
Maxwell's Writings

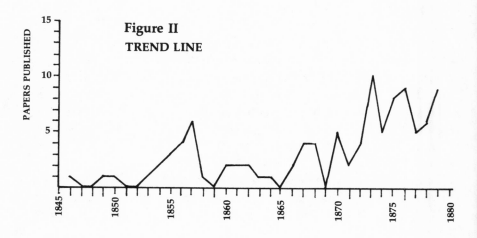

Figure II
TREND LINE

SOURCES: W. D. Niven, ed., *The Scientific Papers of* . . . . Cambridge, 1890. C. W.
F. Everitt, "Maxwell, James Clerk" in *Dictionary of Scientific Biography*.

any terms which do not express relation. The ordinary language about motion and rest does not so completely exclude the notion of their being measured absolutely, but the reason of this is, that in our ordinary language we tacitly assume that the earth is at rest. As our ideas of space and motion become clearer, we come to see how the whole body of dynamical doctrine hangs together in one consistent system. Our primitive notion may have been that to know absolutely where we are, and in what direction we are going, are essential elements of our knowledge as conscious beings. But this notion, though undoubtedly held by many wise men in ancient times, has been gradually dispelled from the minds of students of physics. There are no landmarks in space; one portion of space is exactly like every other portion, so that we cannot tell where we are. We are, as it were, on an unruffled sea, without stars, compass, soundings, wind, or tide, and we cannot tell in what direction we are going. We have no log which we can cast out to take a dead reckoning by; we may compute our rate of motion with respect to the neighbouring bodies, but we do not know how these bodies may be moving in space.[26]

This discussion of the relativity of dynamical knowledge well stated the condition, but not the hopes, of physicists and astronomers around the 1870s. Maxwell's words were also a partial manifestation of the tendency in British philosophy at the time to emphasize its differences with absolutism in all forms, especially those that happened to be German or French. T. H. Green, John Stuart Mill, and Herbert Spencer were among the most vocal exponents of this movement. A classical statement of it may be found in John Stuart Mill's "Examination of Sir William Hamilton's Philosophy," published in 1865. British empiricism was here attacking the tendency among certain mathematicians and mathematical physicists to view the rational mechanics of their deductions as an absolute form of dynamical as well as kinematical knowledge. Maxwell likewise revered mathematics and the certainties of deductive logic, but he was careful to emphasize that the world of dynamical experience could be approached only through empirical or inductive forms of knowledge. In short, Maxwell was a sophisticated natural philosopher whose views of the role of analogical thinking adumbrated many of the attitudes of the twentieth century.[27]

There were other anticipations also of the relativity of position that Maxwell adumbrated. William Kingdon Clifford (1845–79), a mathematical philosopher of high repute in England of the 1870s, was one of the few geometers who took non-Euclidean analyses seriously. He introduced Riemannian surfaces into English mathematics, and his *Common Sense of the Exact Sciences*, completed and published by Karl Pearson (1857–1936) in 1885, had a considerable vogue among those striving to purify mathematical physics of metaphysical content. Clifford's definition of "truth" may be taken as a symbol of the best balance of pride and humility in Victorian science: "Remember, then, that [scientific thought] is the guide of action; that the truth which it arrives at is not that which we can ideally contemplate without error, but that which we may act upon without fear."[28]

Another most unlikely source of criticism of reigning systems of thought came from a German-American philosophical lawyer, judge, and statesman who was highly regarded by Ernst Mach for his theory of cognition. Johann Bernhard Stallo (1823–1900) published a series of articles in *Popular Science* in the 1870s that subsequently became a book entitled *The Concepts and Theories of Modern Physics* (1881). Stallo insisted so strongly on the relativity of all knowledge, ultimately, that he is sometimes seen as a precursor of such secondary effects of relativity theory in the twentieth century as the sociology of knowledge.[29]

During the 1870s, while Maxwell was putting the finishing touches on his kinetic-molecular theory of gases and polishing the discoveries and ideas of Faraday in his own *Treatise*, the world was changing, it seemed, more rapidly than ever before. In 1867, after many vicissitudes, a transatlantic cable came into continuous operation. In 1869, there occurred a double triumph in transportation history that suddenly and dramatically reduced the effective size of the world: the completion of the Suez Canal linking Europe and Asia, and the opening of the American transcontinental railroad linking the Pacific with the Atlantic. Such events may be seen as direct inspiration for Jules Verne's imaginary voyages (*Voyage Autour du Monde en Quatre-vingt Jours* [*Around the World in Eighty Days*] first appeared in *Le Temps* in 1872). These were also the times when the marvel of transatlantic telegraphic communications made the relativity of time common knowledge: because of the cable, events in England could be known in America "before" they happened; that is, local time in both places could be slightly short-circuited.[30]

While Hermann Helmholtz in Berlin and Ludwig Boltzmann in Austria were following paths similar to Maxwell's toward similar conclusions in their theories of heat and light, Maxwell thought of himself as philosophically at odds with all Continental theorists who believed in action-at-a-distance. On the other side of the English Channel the tradition established by the anti-Cartesian, hyper-Newtonian mechanics of Lagrange, Laplace, Poisson, and Gauss had been carried on by Ampère and by the Webers, the Neumanns, and others. Their mathematical formalisms were reducing all electromagnetic phenomena to mere attractions and repulsions exerted across vacuous distances by point-mass particles of matter in motion in empty space. Maxwell's great obsession, like that of Faraday, was to overturn the idea of instantaneous action-at-a-distance. In company with Kelvin and Peter Guthrie Tait (1831–1901), Maxwell felt much more comfortable with the idea of action-by-contact through the use of an insubstantial medium of some sort filling all space. Maxwell's predisposition for continuity meant also that he rejected the notion of a particulate nature for electricity and hoped to prove that light is wave motion in the electromagnetic aether. Maxwell argued this way in beginning Chapter 20, "Electromagnetic Theory of Light," in the 1873 treatise:

781.] In several parts of this treatise an attempt has been made to explain electromagnetic phenomena by means of mechanical action transmitted from one body to another by means of a medium occupying the space between them. The undulatory theory of light also assumes the existence of a medium. We have now to shew that the properties of the electromagnetic medium are identical with those of the luminiferous medium.

To fill all space with a new medium whenever any new phenomenon is to be explained is by no means philosophical, but if the study of two different branches of science has independently suggested the idea of a medium, and if the properties which must be attributed to the medium in order to account for electromagnetic phenomena are of the same kind as those which we attribute to the luminiferous medium in order to account for the phenomena of light, the evidence for the physical existence of the medium will be considerably strengthened.

But the properties of bodies are capable of quantitative measurement. We therefore obtain the numerical value of some property of the medium, such as the velocity with which a disturbance is propagated through it, which can be calculated from electromagnetic experiments, and also observed directly in the case of light. If it should be found that the velocity of propagation of electromagnetic disturbances is the same as the velocity of light, and this not only in air, but in other transparent media, we shall have strong reasons for believing that light is an electromagnetic phenomenon, and the combination of the optical with the electrical evidence will produce a conviction of the reality of the medium similar to that which we obtain, in the case of other kinds of matter, from the combined evidence of the senses.

782.] When light is emitted, a certain amount of energy is expended by the luminous body, and if the light is absorbed by another body, this body becomes heated, shewing that it has received energy from without. During the interval of time after the light left the first body and before it reached the second, it must have existed as energy in the intervening space.

According to the theory of emission, the transmission of energy is effected by the actual transference of light-corpuscles from the luminous to the illuminated body, carrying with them their kinetic energy, together with any other kind of energy of which they may be the receptacles.

According to the theory of undulation, there is a material medium which fills the space between the two bodies, and it is by the action of contiguous parts of this medium that the energy is passed on, from one portion to the next, till it reaches the illuminated body.

The luminiferous medium is therefore, during the passage of light through it, a receptacle of energy. In the undulatory theory, as developed by Huygens, Fresnel, Young, Green, &c., this energy is supposed to be partly potential and partly kinetic. The potential energy is supposed to be due to the distortion of the elementary portions of the medium. We must therefore regard the medium as elastic. The kinetic energy is supposed to be due to the vibratory motion of the medium. We must therefore regard the medium as having a finite density.

In the theory of electricity and magnetism adopted in this treatise, two forms of energy are recognized, the electrostatic and the electrokinetic . . . and these are supposed to have their seat, not merely in the electrified or magnetized bodies, but in every part of the surrounding space, where electric or magnetic force is observed to act. Hence our theory agrees with the undulatory theory in assuming the existence of a medium which is capable of becoming a receptacle of two forms of energy.[31]

Here and elsewhere in Maxwell's writings, particularly in the articles on "Aether" and "Atom" for the ninth edition of the *Encyclopaedia Britannica*, Maxwell showed consummate skill in the use of the conceptual model of the electromagnetic aether as a vehicle for his equations. Less dogmatically than his colleague Kelvin, who sought mechanical explanations as a habitual matter of principle, Maxwell was content to use the aether as an analogy, as a tool for thought without any necessary status in reality. Like the Aristotelians of old, however, Maxwell abhorred the idea of a perfect vacuum, much preferring an imperfect plenum, if a choice were forced.

Maxwell had once attempted an experiment to measure the relative velocity of the earth against the aether of space, and at the end of his life he once again suggested in a letter to American astronomers how this might be done if precise enough optical apparatus could be devised to make the test. In his last years, suffering from cancer while editing the experimental papers of Cavendish, Maxwell again became concerned with the unexplained causes of gravitation and its seemingly instantaneous actions-at-a-distance. He often wondered if someday the electromagnetic approach might be able to solve the causal mystery of gravity, just as he had noted in his letter to Faraday in 1857.

At no time, however, did Maxwell take the position that his theory of electromagnetism was merely his set of differential equations developed to account for and elicit experimental evidence. That was Hertz's statement made for Hertz's own reasons, that is, to make Maxwell palatable to German mathematical physicists. At the very end of the 1873 treatise, Maxwell wrote a section entitled "The Idea of a Medium Cannot Be Got Rid Of," which ended by quoting Torricelli, Galileo's assistant and the father of the mercury barometer and of vacuum studies. He had once remarked that energy

> "is a quintessence of so subtile a nature that it cannot be contained in any vessel except the inmost substance of material things." Hence [Maxwell continued] all these theories lead to the conception of a medium in which the propagation takes place, and if we admit this medium as an hypothesis, I think it ought to occupy a prominent place in our investigations, and that we ought to endeavour to construct a mental representation of all the details of its actions, and this has been my constant aim in this treatise.[32]

Although gravitational attractions remained an intractable example of action-at-a-distance, Maxwell's idea of the electromagnetic field, in which energy could be located in time and its effective forces described as vectors of stress at any point, allowed mathematical physics to venture beyond the old dilemma regarding action-by-contact, as in a direct collision, versus energy transfer by action-at-a-distance. Maxwell seems to have felt subliminally that time had to be involved in all energy exchanges.

Thus the Maxwellian synthesis in retrospect was primarily characterized by the consolidation of the field concept, linking light to electricity and magnetism, the elaboration of the energy concept, and the provision of field equations that could serve to predict a whole new temporal world of electromagnetic phenomena. Mutual relationships between mechanical energy, thermal energy, electrical energy, and magnetic fields thus became domesticated phenomena, susceptible to systematic treatment. Just as studies of actual steam engines and theoretical refrigerators shortly before midcentury had led to the codification of two laws of thermodynamics—the conservation of energy, and the law of entropy or tendency toward disorder—so now students of Maxwell learned to treat electromagnetic energy transfer in terms of conservation and the field. Eventually Maxwell's theory, once translated and elaborated by Helmholtz, Hertz, Oliver Heaviside, Oliver Lodge, G. F. FitzGerald, and others, became trimmed down to four field equations that were indispensable tools for the new age of electricity and later of electronics.[33]

Industrial and instrumental progress in physics and in engineering led toward greater understanding and familiarity in the use of electricity, magnetism, and the electromagnetic spectrum beyond the small range of visible light. This led eventually to a shared reverence among scientists and engineers for James Clerk Maxwell as the father of the theory linking the invisible worlds of radiant energy. Although thermodynamics as a phenomenological theory survived intact into the twentieth century, the changes soon to come in the theory of electrodynamics, atomic physics, and radioactivity were so profound that the Maxwellian synthesis stands as a monument to human understanding of nature only for a relatively narrow range of phenomena. Though also temporally superseded, Maxwell's theory in the form of his field equations still serves well for ordinary applications in the physics needed to explain most of mankind's experience. Only for those who probe atomic nuclei, stellar interiors, and galactic centers has the Maxwellian synthesis failed to satisfy.

NOTES

1. For a broad survey of the pride and prejudices in modern physical science, see Stanley L. Jaki, *The Relevance of Physics* (Chicago: Univ. of Chicago Press, 1967). For a similar study with emphasis on chemistry, see Cecil J. Schneer, *Mind and Matter* (New York: Grove Press, 1969); and for biology, see William Coleman, *Biology in the Nineteenth Century: Problems of Form, Function, and Transformation* (New York: John Wiley &

Sons, 1971), and Elizabeth Gasking, *The Rise of Experimental Biology* (New York: Random House, 1970). For an introduction to the state of scholarship on relativity theory, see Gerald Holton, *Thematic Origins of Scientific Thought: Kepler to Einstein* (Cambridge: Harvard Univ. Press, 1973).

2. The classic contemporary study is, of course, Thomas S. Kuhn, *The Structure of Scientific Revolutions* (Chicago: Univ. of Chicago Press, 1962), but see also the essays in Imre Lakatos and Alan Musgrove, eds., *Criticism and the Growth of Knowledge* (Cambridge: Univ. Press, 1970); Joseph Ben-David, *The Scientists' Role in Society: A Comparative Study* (Englewood Cliffs: Prentice-Hall, 1971); Ronald N. Giere and Richard S. Westfall, eds., *Foundations of Scientific Method: The Nineteenth Century* (Bloomington: Indiana Univ. Press, 1973). For the state-of-the-art in history, see three consecutive issues of *Daedalus* (Journal of the American Academy of Arts and Sciences, vol. 99, no. 4; vol. 100, nos. 1 and 2): *The Making of Modern Science: Biographical Studies*, Gerald Holton, ed. (Fall 1970); *Historical Studies Today*, Felix Gilbert, ed., (Winter, 1971); *The Historian and the World of the 20th Century*, Stephen R. Graubard, ed. (Spring 1971). See also Stephen G. Brush, "Should the History of Science Be Rated X?" *Science* 183 (22 March 1974): 1164–72.

3. Galileo Galilei, *Dialogue Concerning the Two Chief World Systems— Ptolemaic & Copernican*, Stillman Drake, transl. from the 1st ed., Florence, 1632 (Berkeley: Univ. of California Press, 1967), p. 116; cf. pp. 127, 142, 154; and also pp. 114–30, 140–44, 234–42, 369. For biographical details, a first cut may be taken from Trevor I. Williams, ed., *A Biographical Dictionary of Scientists* (New York: Wiley-Interscience, 1969). Better by far are the articles in C. C. Gillispie, gen. ed., *Dictionary of Scientific Biography* [hereafter referred to as *DSB*], 14 vols. (New York: Scribner's, 1970–76).

4. Galileo Galilei, *Dialogues Concerning Two New Sciences*, Henry Crew and Alfonso de Salvio, trans. from the 1st ed., Leyden, 1638 (New York: Macmillan-Dover, 1914) and a new translation by Stillman Drake, *Two New Sciences Including Centers of Gravity & Force of Percussion* (Madison: Univ. of Wisconsin Press, 1974). Stillman Drake also prepared the article on Galilei in *DSB*, V: 237–49.

5. *Sir Isaac Newton's Mathematical Principles of Natural Philosophy and His System of the World*, Florian Cajori, ed., from the English translation by Andrew Motte, London, 1729 (Berkeley: Univ. of California Press, 1946), pp. 8, 12; and now better still, the critical *variorum* edition of Newton's *Principia* is available with I. Bernard Cohen's *Introduction*, 3 vols. (Cambridge: Harvard Univ. Press, 1971–73).

6. Sir Isaac Newton, *Optics: or A Treatise of the Reflections, Refractions, Inflections & Colours of Light*, based on the 4th edn., London, 1730 (New York: Dover, 1952). See also I. Bernard Cohen, ed., *Isaac Newton's Pa-*

pers & Letters on Natural Philosophy (Cambridge: Harvard Univ. Press, 1958); Richard S. Westfall, *Force in Newton's Physics: The Science of Dynamics in the 17th Century* (New York: American Elsevier, 1971); and Westfall's "Newton and the Fudge Factor," *Science* 179 (23 February 1973): 751–58.

7. Arnold Thackray, *Atoms and Powers: An Essay on Newtonian Matter-Theory and the Development of Chemistry* (Cambridge: Harvard Univ. Press, 1970); Robert E. Schofield, *Mechanism and Materialism: British Natural Philosophy in an Age of Reason* (Princeton: Princeton Univ. Press, 1970). For a classic exposition of the development of Newtonian mechanics, see René Dugas, *A History of Mechanics,* trans. J. R. Maddox (Neuchatel: Griffon, 1955).

8. On Bradley, see George Sarton, "Discovery of the Aberration of Light," *ISIS* 16 (November 1931): 233–39; Albert B. Stewart, "The Discovery of Stellar Aberration," *Scientific American* 210 (March 1964): 100–108; Herbert Hall Turner, *Astronomical Discovery* (Berkeley: Univ. of California Press, 1963), from the 1st ed., London, 1904, Chap. 3, pp. 86–120.

9. On Wright, Kant, and Lambert, see *Theories of the Universe—From Babylonian Myth to Modern Science,* Milton K. Munitz, ed. (Glencoe: Free Press, 1957), pp. 225–63; quotations from pp. 254, 257.

10. "William Herschel, 1738–1822" by D. F. Arago, in Bessie Z. Jones, ed., *The Golden Age of Science: Thirty Portraits of the Giants of 19th Century Science by Their Scientific Contemporaries* (New York: Simon and Schuster, 1966) (hereafter cited as Jones, *Golden Age*), pp. 1–26. For an explanation of this whole problem, see William Wallace Campbell, *Stellar Motions* (New Haven: Yale Univ. Press, 1913). See also Michael A. Hoskin, *William Herschel and the Construction of the Heavens* (New York: Norton, 1963), pp. 27–59.

11. William Gilbert, *De Magnete,* P. F. Mottelay, transl., London, 1893, from the original, London, 1600 (New York: Dover, 1958). I. Bernard Cohen, *Franklin and Newton: An Inquiry into Speculative Newtonian Experimental Science and Franklin's Work in Electricity . . .* (Philadelphia: American Philosophical Society, 1956); C. Stewart Gillmor, *Coulomb and the Evolution of Physics and Engineering in 18th-Century France* (Princeton: Princeton Univ. Press, 1971).

12. Luigi Galvani, *Commentary on the Effect of Electricity on Muscular Motion . . .* trans. from the original, Modena, 1792, by Robert M. Green (Cambridge, Mass.: E. Licht, 1953); Georgio de Santillana, "Alessandro Volta," *Scientific American* 212 (January 1965): 82–91; Bern Dibner, *Oersted and the Discovery of Electromagnetism* (New York: Blaisdell, 1962).

13. For biographical history of the great mathematicians, see E. T. Bell, *Men of Mathematics* (New York: Simon and Schuster, 1937); for Euler see chap. 9, "Analysis Incarnate," pp. 139–52.

14. For Cauchy, see Bell, *Men of Mathematics,* chap. 15, "Mathematics and

Windmills," pp. 270–93, and for Gauss, chap. 14, "The Prince of Mathematicians," pp. 218–69.

15. For the interrelations of mathematics, mechanics, astronomy and optics, electricity and magnetism, especially from a French perspective, see René Taton, ed., *History of Science*, trans., A. J. Pomerans, Vol. III, *Science in the Nineteenth Century* (New York: Basic Books, 1965), pp. 7–234.

16. Morton L. Schagrin, "Resistance to Ohm's Law," *American Journal of Physics* 31 (July 1963), 536–47.

17. Ibid., pp. 185–207; Stephen G. Brush, "The Wave Theory of Heat," *British Journal for the History of Science* 5 (1970): 145–67. See also Brush's book in this series, *The Temperature of History: Phases of Science and Culture in the Nineteenth Century* (New York: Burt Franklin, 1978).

18. Nathan Reingold, ed., *The Papers of Joseph Henry*, vol. I, *The Albany Years* [1797–1832]. L. Pearce Williams, *Michael Faraday: A Biography* (New York: Basic Books, 1965); Joseph Agassi, *Faraday as a Natural Philosopher* (Chicago: Univ. of Chicago Press, 1971).

19. "Lord Kelvin (William Thomson), 1824–1907," by S. P. Thompson, in Jones, *Golden Age*, pp. 466–90. See also W. Thomson's 1871 Presidential Address to the British Association (BAAS) meeting in Edinburgh, available in George Basalla, William Coleman, Robert H. Kargon, eds., *Victorian Science* (Garden City: Doubleday Anchor Original, 1970), pp. 98–128.

20. Maxwell's letter to Faraday, 9 November 1857, is reprinted *in extenso* in L. Pearce Williams, *The Origins of Field Theory* (New York: Random House, 1966), pp. 117–20. For an excellent recent study that rates Faraday higher than any of his successors, see William Berkson, *Fields of Force: The Development of a World-View from Faraday to Einstein* (New York: Halstead Press, 1974).

21. Joseph Larmor, ed., *Origins of Clerk Maxwell's Electric Ideas* as described in familiar letters to William Thomson (Cambridge: Univ. Press, 1937).

22. D. K. C. MacDonald, *Faraday, Maxwell and Kelvin* (Garden City: Doubleday Anchor Book, 1964). Rollo Appleyard, *Pioneers of Electrical Communication* (London: Macmillan, 1930), on Maxwell, pp. 3–30. For an anecdotal history, see Herbert W. Meyer, *A History of Electricity and Magnetism* (Cambridge: MIT Press, 1971).

23. Lewis Campbell and William Garnett, *The Life of James Clerk Maxwell . . .* (London: Macmillan, 1882); R. T. Glazebrook, *James Clerk Maxwell and Modern Physics* (New York: Macmillan, 1905). See also the excellent sketch by C. W. F. Everitt in Gillispie, *DSB*, Vol. IX, pp. 198–230, which is available in expanded form as C. W. F. Everitt, *James Clerk Maxwell: Physicist and Natural Philosopher* (New York: Charles Scribner's Sons, 1975).

24. See the discussion in Kenneth F. Schaffner, *Nineteenth-Century Aether Theories* (Oxford: Pergamon Press, 1972), pp. 78–84.

25. See Alfred M. Bork, ed., "Foundations of Electromagnetic Theory—Maxwell," forthcoming in *Sources of Science* (Johnson Reprint series). See also Michael J. Crowe, *A History of Vector Analysis: The Evolution of the Idea of a Vectorial System* (Notre Dame, Ind.: Univ. of Notre Dame Press, 1967).

26. James Clerk Maxwell, *Matter in Motion*, ed. Joseph Larmor, 1st ed., London, 1876 (New York: Dover, n.d.), p. 81. On "Relativity of Knowledge," see *Encyclopaedia Britannica*, 11th ed., XXIII: 58–59.

27. Mary B. Hesse, *Forces and Fields: The Concept of Action at a Distance in the History of Physics* (Totowa, N.J.: Littlefield, Adams, 1965). Cf. Mary B. Hesse, *Models and Analogies in Science* (Notre Dame, Ind.: Univ. of Notre Dame Press, 1966). For an astronomer's prevision of the relativity of simultaneity (albeit a naïve preview), see Richard A. Proctor, *Other Worlds than Ours: The Plurality of Worlds Studied Under the Light of Recent Scientific Researches*, Ira Remsen, ed., original ed. c. 1870s (New York: J. A. Hill, 1904), p. 206.

28. William Kingdon Clifford, *The Common Sense of the Exact Sciences*, ed. Karl Pearson, orig. edn., London, 1885; rev. ed., ed. James R. Newman (New York: Dover, 1955), p. ii.

29. Inveighing against a series of metaphysical errors, J. B. Stallo wrote: "There is no absolute material quality, no absolute material substance, no absolute physical unit, no absolutely simple physical entity, no absolute physical constant, no absolute standard, either of quantity or quality, no absolute motion, no absolute rest, no absolute time, no absolute space." Arguing for "the relativity of all objective reality" as early as 1873, Stallo insisted that "the fact that everything is, in its manifest existence, but a group of relations and reactions at once accounts for Nature's inherent teleology." See J. B. Stallo, *The Concepts and Theories of Modern Physics*, ed. Percy W. Bridgman, from orig. edns. in 1881, 1884, 1888 (Cambridge: Belknap of Harvard Univ. Press, 1960), pp. 201, 202.

30. On the relativity of local times, see Robert Routledge, *Discoveries and Inventions of the Nineteenth Century* (London: Routledge & Sons, 1879), p. 431. For a guide to the history of technology, see Melvin Kranzberg and Carroll Pursell, eds., *Technology in Western Civilization*, 2 vols. (New York: Oxford Univ. Press, 1967).

31. James Clerk Maxwell, *A Treatise on Electricity and Magnetism*, 2 vols. (Oxford: Clarendon Press, 1873), II: 383, 384.

32. Ibid., II: 403.

33. Cf. three commemorative sets of essays on his stature: Sir J. J. Thomson, et al., *James Clerk Maxwell . . . 1831–1931: Essays* (Cambridge: Univ. Press, 1931); R. L. Smith-Rose, *James Clerk Maxwell, F.R.S., 1831–1879* (London: Longmans, Green, 1948); C. Domb, ed., *Clerk Maxwell and Modern Science: Six Lectures* (London: Athlone Press, 1963).

Chapter II

Analyses of Radiation
(c. 1880s)

I see very well that . . . the mathematical language of Clerk Maxwell's theory expresses the laws of the phenomena very simply and very truly. . . . But I confess I should really be at a loss to explain . . . what he considers a quantity of electricity and why such a quantity is constant, like that of a substance.

—HERMANN VON HELMHOLTZ, 1881

I want to understand light as well as I can without introducing things that we understand even less of. That is why I take plain dynamics.

—SIR WILLIAM THOMSON
[LORD KELVIN], 1884

I have never been fully satisfied with the results of my Potsdam experiment, even taking into account the correction which Lorentz points out.

—ALBERT A. MICHELSON TO
LORD RAYLEIGH, 6 March 1887

The experiments described appear to me, at any rate, eminently adapted to remove any doubt as to the identity of light, radiant heat, and electromagnetic wave-motion.

—HEINRICH HERTZ, 1888

Albert Einstein was born on 14 March 1879 in Ulm, Würtemberg. James Clerk Maxwell died later that year, on 5 November, in Cambridge, England. The few months' overlap of their lives symbolizes more than mere life spans; Maxwell had formulated the main problems during a new age of electricity that Einstein was destined to transform during a new age of electronics.

The acceleration of history from 1879, when Maxwell died,

through 1891, when Einstein began to know himself, was never more obvious. And yet the slowly maturing young son of Ulm, now moved to Munich, was fated to answer and ask even more problems and questions. In looking at the 1880s, the first decade of his times, we ought to be able to learn more trustworthy and significant things about his wider heritage than he himself could ever know.

We shall first look at more background for the experimental physics that characterized this decade. Helmholtz's dual role as Maxwell's mediator and as mentor for Hertz's generation of electrodynamicists will be described before we delve deeper into cultural differences and certain personalities who tended to set national styles in science. Then we shall see how, in the 1870s, certain experiences happened to shift opinions on radiation. Analyses of radiation during the 1880s seemed to polarize alternately, with some Germans believing in aether waves and some British students of the subject favoring discrete aether particles.

Although confusion was rampant with all the disparate evidence accumulating from radiation studies, two experimentalists emerged from the decade with results so startling that their names eventually came to be synonymous with the major developments of the period. The actual characters of Albert Abraham Michelson (1852–1931) and Heinrich Rudolf Hertz (1857–94) were quite different from the reputations that their works acquired in later years, but we shall see how their respective "aether-drift" and "aether wave" experiments were interpreted as contradictory evidence against and for, respectively, the electromagnetic aether. The sterility of the former experiments and the fertility of the latter in terms of practical applications contributed more than pure scientists care to admit to the growing demand for an electromagnetic world-view at the beginning of the twentieth century.

The 1880s were an extraordinary period for physical science, even though the next several decades were progressively so much more so that in retrospect the 1880s have seemed mild by comparison. If we try to immerse ourselves, with a prospective attitude, imaginatively and sequentially into the years after 1880, we find that the achievements of the then recent past, the challenges of the moment, and the expectations for the future impressively demonstrate the progressive feelings and attitudes held by most Victorian intellectuals. One of the most profound set of questions, amenable to experimental and theoretical advances in knowledge, concerned the exact relationships between heat, light, sound, electricity, and magnetism.

Throughout western Europe after the unification of Italy and Germany and the growth of imperial hegemonies in Africa, Asia, and Oceania, there were economic, industrial, and military demands for applying science and technology to world markets, to communication and transportation systems, and to the basic needs for improving food supplies, as well as clothing and housing materials. Thus chemistry, the science of materials, became the glamour science of the age. Physicists often took pride in practicing a more purely intellectual profession designed to answer ancient philosophical questions relating to the nature of heat, light, electricity, magnetism, matter, and even mind. Gustav Fechner (1801–87) and Ernst Mach (1838–1916), for example, set themselves certain goals for developing a marriage of psychology and physics, or psychophysics.[1]

Perhaps the most prevalent theses of the decade among philosophical physicists were the notions that progress depended on positive knowledge based on exhaustive analyses and on acquisition of better data regarding all sorts of energy transfer, especially radiation. Inherited confidence in the concepts of matter and motion, time and space, force and energy were being undermined in many ways, not least of which was by ever more precise measurements in astronomy, optics, acoustics, thermodynamics, and electrodynamics. The success of the undulatory theory of light and the promise of Maxwell's theory linking light and electromagnetism were such as to inspire much confidence in the possibility of reducing the problem of matter—What is it, ultimately?—to a problem of energy or vortex motions of ultramundane portions of the ubiquitous aether. Such ancient concerns enjoyed a revival once again during the 1880s.

This ethereal world-view was always entertained cautiously by serious thinkers, of course, because there were many difficulties with radiant light and heat phenomena that seemed intractable. The most astute investigators of these problems, however, gradually built up a consensus over the decade that recognized the need to move from analogous to analytical thinking and from materialistic to mechanistic hypotheses. It is our purpose in this chapter to see this consensus develop.

Basic to understanding the scientific debates over the microstructure and macrostructure of the universe toward the turn of the century is an appreciation of the growth of tools and techniques for dealing with—practically—nothing and everything! This hyperbole hardly exaggerates what the nineteenth century had already achieved by the year of Maxwell's death and Einstein's birth. Powerful new instruments existed for probing evacuated spaces and astronomical distances. There was a central nexus between vacuum and plenum studies around 1880. Microphysics was learning most valuable lessons from studying what happens in small chambers where *almost* nothing exists. Macrophysics was learning much from studies that focused on synoptic views of large-scale spaces—the solar system, the whole earth, and our whole galactic neighborhood. The only obvious link, light, had already become, through spectroscopy, photometry, and photography, the most reliable guide to knowledge about stars and atoms. Hopes were justifiably high that such linkages between terrestrial chemical discoveries and celestial astrophysics would continue to prove highly fruitful.[2]

The paradox represented by the assertion that everything is nothing and nothing everything comes close to expressing the main problem with the ethereal world-view. Similarly, the paradox that motion can be physically defined only in relation to matter—and vice versa—came to be a favorite metaphysical exercise toward the end of the century. Indeed, with the publication of Ernst Mach's influential book *The Science of Mechanics* (first issued in 1883 then revised and republished seven times before 1912), the "effort to get rid of metaphysical obscurities," as Mach first phrased his purpose, became a conscious concern of most scientists who also considered themselves philosophers. Mach's own philosophy of science was rooted in a strong sense of history and a profoundly critical approach to primary assumptions. As we shall see, Mach's gospel of epistemological relativity, his thorough critique of Newtonian mechanics, especially of the ideas of absolute time and absolute space, and his ongoing influence in shaking the dogmatic faith of young Europeans make him a major transitional figure—like Helmholtz—between Maxwell and Einstein.

Crookes's Radiometer and Goldstein's Rays

Leading into some central concerns of the 1880s were two episodes in the history of heat, light, and electromagnetism that could not have happened without the technology to create glass vacuum tubes of a high order. The first of these episodes concerns a famous controversy over a curious little device called a "radiometer," in which furor Maxwell played a leading role. The second episode relates to the colorful studies of electric discharges in rarefied gases, which began in earnest the year Maxwell died.

William Crookes (1832–1919), who had learned chemistry from A. W. Von Hofmann in London around midcentury, became famous after his discovery of the new element thallium in 1861. His exploitation of spectroscopic techniques and physical balances operated in a vacuum established his reputation as one of the most meticulous British chemists. But Crookes moved toward physics in the 1870s while investigating certain anomalies related to weighing hot and cold samples in ever better vacua.

Heinrich Geissler (1814–79) had perfected a mercury-piston vacuum pump in 1855 with which he and Julius Plucker (1801–68) had started studies of the luminous glows radiated by different gases at low pressures. These colorful aurorae appeared only inside blown-glass globes through which a current was introduced by electrodes. In 1858, Plucker had reported that striated purple glows sputtered and could be deflected in a strong magnetic field. J. W. Hittorf (1824–1914), making use of H. J. P. Sprengel's (1834–1906) improved mechanized vacuum pump, collaborated with Plucker in reaching new lows of internal pressure and discovering two distinct types of discharge-tube spectra. Hittorf's basic paper on cathode discharges (1869), reporting sharp shadows and strictly rectilinear transmissions, was based on residual pressures below 2 mm and down to ⅓ mm of a mercury barometer.[3]

Crookes took full advantage of these parallel developments for his own researches. In 1874, he published notices of a series of experiments on "The Repulsion Resulting from Radiation." Among the ingenious devices he described was a "light-mill" called a "radiometer," which created an immediate sensation. This device is still a favorite attention-arrester in jewelers' windows and toy shops throughout the world. Crookes argued that the radiometer demonstrated and could measure the force of radiation pressure. The delicate cruciform arms, holding vanes blackened on one side, silvered on the other, and balanced on a needle point in a well-evacuated globe, would spin at speeds depending, it seemed, only on the light intensity.

Crookes's radiometer was well publicized, fairly easy to duplicate, and delightfully fascinating. It quickly became a demonstrative toy as well as the focus of a theoretical debate over its behavior. Maxwell had predicted that light pressure of slight degree might someday be measurable. Although Crookes was unaware of this when developing his light-mill, experiment and theory were here brought into immediate conflict. The trouble was that Crookes's radiometer always spins with the black faces receding from the light source; this rotation is opposite to what Maxwell and the other wave theorists expected if radiation pressure were the real cause of the motion.

The fact that the bright surface reflects radiant heat and the black side absorbs it, converting light directly into motion, seemed to involve a fundamental contradiction of the laws of mechanics. Newton's third law would be violated and Crookes's own earlier experiments on hot and cold bodies in vacuum balances would be contradicted if the cause of motion were attributed to light pressure alone. Arthur Schuster (1851–1934) devised a crucial experiment in 1876 by suspending the whole case of a radiometer very delicately in order to see which way it would turn when incident light struck the vanes. The verdict was clearly in favor of Newton's third law. So Crookes came to accept a simple kinetic-molecular explanation for this light-mill. Typically it operated at a pressure of .1 mm of mercury but would still work at "low" pressures as high as 35 mm. Physicists all over Europe, but especially Maxwell, Osborne Reynolds (1842–1912), P. G. Tait, James Dewar, Johann Zollner, and S. T. Preston, began proposing better explanations. As a result, the idea of the mean free path for agitated molecules in the kinetic theory of gases was thoroughly reworked. Maxwell and Reynolds independently provided solutions to the paradox by showing with sophisticated mathematics how an unbalanced force is created by a slipstream near the edges of the blackened and therefore heated sides of the vane, where the heat flow in the gas is not uniform.[4]

Meanwhile, by 1878 Crookes had profited from all this plus more experiments to develop a new form of this device, the cup radiometer. This model with polished insides and darkened outsides would spin the "wrong" way too, with its cups forward, opposite to the direction expected from the analogy of an anemometer. This remarkable episode not only demonstrated the importance of residual phenomena in supposed vacua; it also laid the foundations for rarefied gas dynamics; it led Reynolds into his classic analyses on fluid dynamics and turbulent flow; and it caused Crookes to intensify his experimental work with vacuum-discharge phenomena.

In 1879, Crookes delivered the Bakerian Lecture "On the Illumination of Lines of Electrical Pressure, and the Trajectory of Molecules" wherein he demonstrated and described some truly marvelous vacuum tubes made by his friend and assistant, C. H. Gimingham, an artist at glass-blowing and manipulation of glass. By now Crookes's studies of cathode discharges were based on tubes so empty of residual air as to be measured in new units: 30 M down to 14 M, where M represents a millionth of a millimeter. Also by now Crookes was echoing Faraday's belief in "radiant matter." With clever transparent apparatus to demonstrate his ideas, Crookes challenged his peers with experimental tests and with his daring conclusions about a fourth, or an "ultragaseous," state of matter:

> The modern idea of the gaseous state of matter is based upon the supposition that a given space of the capacity of, say, a cubic centimeter contains millions of millions of molecules in rapid motion in all directions, each having millions of encounters in a second. In such a case the length of the mean free path of the molecules is excessively small as compared with the dimensions of the vessel, and properties are observed which constitute the ordinary gaseous state of matter, and which depend upon constant collisions. But by great rarefaction the free path may be made so long that the hits in a given time are negligible in comparison to the misses, in

which case the average molecule is allowed to obey its own motions or laws without interference; and if the mean free path is comparable to the dimensions of the vessel, the properties which constitute gaseity are reduced to a minimum, and the matter becomes exalted to an ultra-gaseous or molecular state, in which the very decided but hitherto masked properties now under investigation come into play.

The phenomena in these exhausted tubes reveal to physical science a new world—a world where matter may exist in a fourth state, where the corpuscular theory of light may be true, and where light does not always move in straight lines, but where we can never enter, and with which we must be content to observe and experiment from the outside.[5]

Meanwhile, Eugen Goldstein (1850–1930) in Berlin, another of Helmholtz's many students, was hard at work on similar experiments with what he first called "cathode rays." In some ways he anticipated Crookes's work, especially with measurements of exact rectilinear propagation, concave cathodes for focusing the rays, and tests to show that the rays were independent of the type of material conducters used in electrodes. But Goldstein, like his friends and colleagues E. E. G. Wiedemann (1852–1938) and Heinrich Hertz, strongly advocated the wave theory. Whereas Crookes was arguing for "molecular rays" and being supported by G. Johnstone Stoney (1826–1911), he was opposed by Arthur Schuster and Peter Tait, among others. In 1874, Stoney had first proposed the term "electrine" for an absolute unit of electricity that could be used for explaining chemical bonds or valency. But this was forgotten till Helmholtz resurrected the idea of atomized energy in 1881. The wave-versus-particle controversy grew stronger after 1886 when Goldstein announced his discovery of *Kanalstrahlen*, or "canal rays," as a counterpart phenomenon to cathode rays. The blue or green glow customarily encountered in cathode displays was shown to be accompanied by other discharges from the cathode. Distal from the anode and behind a perforated cathode, Goldstein, using elongated vacuum tubes, discovered columns of yellow light at the back of the cathode. This was a quite separate phenomenon, less impressive visually but equally interesting because these rays were positive rather than negative. Goldstein and Hertz were upholding a minority view in Germany, just as Stoney and Crookes were in the minority in Britain. But they all achieved such eminence for other reasons that their influence on informed opinions was heavy.[6]

Throughout the 1880s, Crookes delved deeper into the mysteries of the rare earths, especially yttrium. He produced better vacuum pumps, more elaborate tubes and instruments. At the same time Thomas A. Edison and Joseph W. Swan were perfecting their incandescent lamps. Edison's patent and Swan's priority in filament studies were reconciled in Britain by a commercial union in 1883 to exploit the "Ediswan" lighting and power system. "Crookes tubes" had played a role in advance of these developments just as the so-called Edison effect would later become important in thermionics. Edison noticed that a cathode discharge could be regulated like a valve. This he attributed to an "etheric force."

In time Edison's discovery would become the basis for the diode, and later the triode, but not before Goldstein's "cathode rays" would become electrons; and Crookes's radiometer would be superseded by a better one made by P. N. Lebedev (1866–1912) in Russia. In 1899, Lebedev would be first to verify Maxwell's radiation pressure by other means. By that time Crookes would be Sir William (knighted in 1897), still learning and adapting to a changing world. One of his avocations was investigating so-called psychic phenomena. Emanations, radiations, and aether-waves of all sorts by then seemed to offer limitless possibilities—almost.

Helmholtz as Modifier of Maxwell

Although the Maxwellian synthesis seemed stillborn, because optics and electromagnetism were not fully reconciled and incorporated by 1880, the decade to follow witnessed a resurgence of interest in analyzing the mysteries of radiant heat, light, and energy. Rays and waves of all types were being investigated as never before, largely because of Maxwell's suggestions. No one appreciated the need for this more than the leading scientist of the leading scientific nation, Germany.

Hermann Ludwig Ferdinand Helmholtz (1821–94) was widely recognized by 1880 as one of the foremost scholar-scientists of the nineteenth century. The son of a Potsdam teacher of philology and philosophy, young Helmholtz first became a surgeon in the Prussian Army and then moved through various physiological professorships until called to the chair of physics in Berlin in 1871. Along with J. R. Mayer, J. P. Joule, and Lord Kelvin around midcentury, Helmholtz had played a major role in establishing the principle of the conservation of "force," later called energy. His studies in physiology, optics, acoustics, chemistry, mathematics, electricity, magnetism, meteorology, and theoretical mechanics made him seem a most remarkable renaissance man. His great work on *Physiological Optics,* finished after a decade in 1866, was paralleled by another magnum opus in physiological acoustics, *The Sensations of Tones,* published in 1863. These works had all prompted his call to Berlin. There, in the 1870s, Helmholtz began to concentrate more on hydrodynamics and electrodynamics. Receptive to the electromagnetic theories of Maxwell, Helmholtz invited some of his best students to try to verify, refute, or improve upon the Maxwellian assumptions and deductions. Before 1870, Helmholtz had accepted to a large degree Weber's doctrine regarding electricity, but his doubts grew after studying Maxwell's critique of action-at-a-distance. Soon Helmholtz was encouraging all his best students to make various tests for electric and magnetic energy transfers.

Once he had moved from medicine into physiology and on into physics, Helmholtz's most influential contribution after the 1847 memoir on "The Conservation of Force" was his essay of 1858, "On the Integrals of the Hydrodynamic Equations Which Express Vortex Motion." Mathematicians appreciated the elegance of his equations, and physicists used his insights to advance not only hydrodynamics, meteorology, and oceanography but also theories of the vortex atom. Here Helmholtz laid the foundations for mathematical analyses of minute spheres or rings of a fluid that may move as a whole in some definite direction while also changing shape aand rotating around their own axes. In an ideal, frictionless, and incompressible liquid, Helmholtz proved that vortex motion could be neither created nor destroyed; thus, trying to match the mathematical properties of vortexes to the astounding experimental behavior of smoke rings became a challenge of high order. Helmholtz and Kelvin first met and became friends in 1852. Almost yearly thereafter, these two polymaths were to enjoy each other's company, and Sir William (as Kelvin was then known) grew ever more intrigued with the possibilities for a vortex theory of the atom (1867) until late in life. (After elevation to the peerage in 1892, Lord Kelvin confessed to an astonished audience that his whole life had been characterized by his "failure" to work out a definitive vortex atom theory.)[7]

On 5 April 1881, Hermann Helmholtz (the "von" was to be added the next year, when Kaiser Wilhelm I raised him to the rank of hereditary nobility) delivered the Faraday Lecture before the fellows of the Chemical Society in London. Speaking on "The Modern Development of Faraday's Conception of Electricity," Helmholtz paid tribute first to Faraday's discoveries and then to his method of trying to understand "only facts, with the least possible use of hypothetical substances and forces." At the same time, Helmholtz praised the recently departed Clerk Maxwell for giving the mathematical interpretation of Faraday's conceptions and for reconstructing "in the normal methods of science the great building" that Faraday had planned.

> Nobody can deny that this new theory of electricity and magnetism originated by Faraday and developed by Maxwell, is in itself consistent, is in perfect and exact harmony with all the known facts of experience, and does not contradict any one of the general axioms of dynamics, which have been hitherto considered the fundamental truths of all natural science because they have been found valid, without any exception, in all known processes of nature. A confirmation of great importance was given to this theory by the circumstance, demonstrated by Clerk Maxwell, that the qualities which it must attribute to the imponderable medium filling space are able to produce and sustain magnetic and electric oscillations, propagating like waves and with a velocity exactly equal to that of light. Several parts even of the theory of light are deduced with less difficulty from this new theory than from the well-

known undulatory theory of Huygens, which ascribes to the luminiferous aether the qualities of a rigid elastic body.[8]

After discussing competing electromagnetic theories put forward by Wilhelm Weber, Riemann, and Clausius, Helmholtz expressed his own debts to Kelvin and Tait for their *Treatise on Natural Philosophy* before confessing his belief that the Faraday-Maxwell theory would shortly attain "general assent." Helmholtz reported that he had gone beyond Maxwell's theory for closed circuits to find that the phenomena of "open circuits" were also in full accord with Maxwell's theory. He cited three examples: (1) the oscillatory discharge of a condenser through a coil of wire, (2) his own experiments on electromagnetically induced charges of a rotating condenser, and (3) Rowland's experiments on the electromagnetic effect of a rotating charged disk. Although Helmholtz still preferred a "two-fluid" theory of electricity, he was also sympathetic toward the "field" concept, though avoiding the word, and he strongly recommended ignoring direct forces working at a distance, for they seemed either not to exist or to be negligible "when compared with the tensions and pressures of the dielectric medium":

> I see very well that this assumption of two imponderable fluids of opposite qualities is a rather complicated and artificial machinery and that the mathematical language of Clerk Maxwell's theory expresses the laws of the phenomena very simply and very truly with a much smaller number of hypothetical implications. But I confess I should really be at a loss to explain, without the use of mathematical formulas, what he considers a quantity of electricity and why such a quantity is constant, like that of a substance. The original, old notion of substance is not at all identical with that of matter. It signifies, indeed, that which behind the changing phenomena lasts as invariable, which can be neither generated nor destroyed, and in this oldest sense of the word we may really call the two electricities substances.

As might be expected before a chemical society, Helmholtz concen-

trated on electrochemical developments and on the growth of confidence in Faraday's "ions" and in electrical theory applied to
chemical atoms. The most often quoted part of his address is the
following:

> Now, the most startling result of Faraday's law is perhaps
> this. If we accept the hypothesis that the elementary sub-
> stances are composed of atoms, we cannot avoid concluding
> that electricity also, positive as well as negative, is divided
> into definite elementary portions, which behave like atoms of
> electricity. As long as it moves about in the electrolytic liquid,
> each ion remains united with its electric equivalent or equiva-
> lents. At the surface of the electrodes, decomposition can take
> place if there is sufficient electromotive force, and then the
> ions give off their electric charges and become electrically neu-
> tral.[9]

This, perhaps the most surprising portion of Helmholtz's lecture,
caused great excitement among his British auditors. For here was a
statesman of science with great authority who was interpreting the
need to hold Maxwellian continuity and Daltonian discreteness in
suspended animation. Speaking as though Maxwell were not dead
and as if electromagnetic forces at both the molecular and the
atomic levels might ultimately be reduced to electrical charges,
Helmholtz concluded:

> I think the facts leave no doubt that the very mightiest among
> the chemical forces are of electric origin. The atoms cling to
> their electric charges, and opposite electric charges cling to
> each other; but I do not suppose other molecular forces are
> excluded, working directly from atom to atom. . . . The fact
> that even elementary substances, with few exceptions, have
> molecules composed of two atoms, makes it probable that
> even in these cases electric neutralization is produced by the
> combination of two atoms, each charged with its full electric
> equivalent, not by neutralization of every single unit of affin-
> ity.

Students of Helmholtz

Although everyone, whether he realized it or not, was in some sense a student of Helmholtz after 1880, so far-reaching were his influences beyond Bismarck's Germany that I shall single out three young men—Rowland, Michelson, and Hertz—to examine for their transmissions of his style and wisdom.

Henry Augustus Rowland (1848–1901), from the new Johns Hopkins University in Baltimore, had managed by persistence to interest Helmholtz in a proposal to run an electrical convection experiment under his supervision. That test consisted primarily of a rapidly rotating (60 rps) charged disk of ebonite (or vulcanite), 21 cm in diameter, gilded and radially grooved, with a sensitive astatic needle system to decide between the autonomous electric fluid theory and the Faraday-Maxwell interaction or displacement theory of electromagnetism. Both Maxwell and Helmholtz had encouraged Rowland to proceed, and the test evolved into one for "open circuits," the ratio of units, and "convection/conduction streams." With Helmholtz's patronage, Rowland achieved some fame in 1876 for having given some "positive proof" that moving electrified bodies are also electromagnetically effective. But Helmholtz interpreted Rowland's results as inconclusive in settling the contest between Maxwell's idea of a displacement current and his own theory of electric potential, which emphasized the idea of dielectric polarization. In 1879, Rowland, back in Baltimore, and his student Edwin H. Hall (1855–1938) reported another new magnetic effect of electricity. From then on Rowland abandoned the fluid theory of electromagnetism and moved steadily toward an aether model, praising Maxwell's interactive theory above all others.[10]

Albert A. Michelson arrived in Europe for postgraduate studies during 1880. He came to Helmholtz's laboratory in the wake of Rowland and while Heinrich Hertz was still a demonstrator there. But the slightly more mature Michelson divided his time among several European universities, seeking support for an important test he had in mind. Michelson, a U.S. naval officer in mufti, had invented a simple device that he hoped might serve as a combined speedometer and compass for measuring the relative motion of spaceship Earth as it hurtles through the universal ocean of luminiferous aether.

The need for such a device to measure a possible second-order effect had been suggested by Maxwell, and independently the components had been devised by Jules Jamin, Eleuthere Mascart, and Alfred Cornu. These French (and other German) opticians in the mid-nineteenth century had been working on interferential refractometers of various sorts for measuring indices of refraction in various kinds of translucent materials. Applications of such optical work for the German dye industry, for the growing carbide and natural-gas illumination industry, and for quality-control work in all sorts of chemical processes were growing more important daily. But the purely scientific, optical, photographic, and astronomical applications of refractometry were as yet not fully exploited. Michelson had the idea of building an interferential refractometer with optical paths of equal lengths at right angles to each other. He would put this cross of light on a rotating base so that the speed of light in the direction of and at right angles to the supposed resultant vector of Earth's motion through interstellar space could be compared. Later called simply an interferometer, Michelson's apparatus, modified into other designs, became one of the most important scientific instruments ever made for measuring extreme lengths, both large and small. And although it was used in numerous other ways as well as in one of the most celebrated experiments of modern physics, it never fulfilled the *original* intentions of its designer.

Albert Michelson had been born in 1852 in Strzelno, a village in dispute between Prussia and Poland. His parents, Samuel and Rosalie Michelson, emigrated from Europe taking their first-born son with them to America, first to New York City and shortly thereafter to San Francisco by way of an Isthmian crossing. Young Albert grew up during Gold Rush days in central California, but when his family followed the Silver Rush to Virginia City, Nevada, Albert was left to board with a private schoolmaster in San Francisco. His early aptitudes in science and art were encouraged, and upon graduation young Michelson (having taken the middle name Abraham after the assassination of Lincoln) sought to enter the United States Naval Academy through a congressional appointment from Nevada. Unsuccessful in that, Michelson boarded the newly completed transcontinental railroad for a journey direct to the White House to plead his case. President Ulysses S. Grant, touched by such ambition and initiative, arranged for an extralegal appointment to the Naval Academy, one of many that year.

Michelson did not excel in seamanship at Annapolis, but he did graduate first in physics and chemistry. Cruises on sailing vessels and courses in navigation and steam engineering, together with actual ship-handling at sea, made Michelson acutely aware of disjunctions between theory and practice and of problems of measuring relative motion. Michelson's technical education continued after graduation when, after short tours of sea duty, he was reassigned to Annapolis to teach physical sciences.[11]

In 1878, he first approached Professor Simon Newcomb, editor of the U.S. Navy's *Nautical Almanac* and America's leading astronomer at the time. Michelson had been fascinated by laboratory demonstrations of optical phenomena and had done some research on the means used by Foucault to measure the speed of light with revolving mirrors. A gift of $2,000 from his wealthy new father-in-law had given Michelson the working capital to build an apparatus improved in several ways over Foucault's spinning-mirror technique and the latest work of Alfred Cornu. Simon Newcomb, impressed with the new apparatus as well as with the precision and meticulous care of the young instructor's measurements, invited Michelson to publish his results in the *American Journal of Science*, to become a part of his staff, and to continue work to improve the accuracy of measurements of the velocity of light.

In these circumstances, Michelson became a part of the small scientific community around Washington, D.C., and Baltimore. Soon he gained the privilege of a leave of absence for postgraduate study in Europe, going first to Paris, then to Berlin and Heidelberg to study, respectively, under Professors Cornu, Helmholtz, and Quincke. Aware of one of Maxwell's last letters to Newcomb's colleague D. P. Todd of the *American Ephemeris*, which had advised of the need for a second-order (that is, round-trip) measurement to test whether the velocity of light varied in different directions owing to the motion of Earth through the aether, Michelson went to Europe with an important problem in mind. It just might be possible to build an optical current-meter that could detect not only Earth's *orbital* motion but also its *absolute* motion, as Newton had conceived it, if the luminiferous aether could be made to show a relative wind.

In November, Michelson showed Helmholtz his design for an instrument to measure Earth's absolute motion through space, or at least its relative motion with respect to the luminiferous aether. Helmholtz listened carefully as Michelson described his plan for a way to test Maxwell's suggestion that Earth's motion (or at least the velocity of light in different directions) be measured to see also if received theories of astronomical aberration were correct. Maxwell had thought second-order measurements would be virtually insensible because so small was the ratio (one part in a hundred million) for Earth's orbital velocity compared to the velocity of light. But here was Michelson saying that a small brass instrument with meter-length arms at right angles might do the trick. He proposed to race two pencils from a split-beam of light along mutually perpendicular paths, then recombine the beam for an interference pattern of fringes that should shift laterally if one pencil were retarded relative to the other. The idea was marvelously simple. Helmholtz heartily approved the trial, but warned of the extreme criticality of keeping a constant temperature. The tiniest difference in heating could destroy the trustworthiness of any results.

Michelson found in Alelxander Graham Bell another patron to pay for the manufacture of the instrument he designed. The Berlin instrument-makers Schmidt und Haensch handcrafted the brass base, arms, optical parts, and turnstile. A small lamp, a little telescope with a micrometer eyepiece, and a half-silvered beam-splitter were the critical parts. The device was finished in a couple of months.

Early in 1881, Michelson tested his balanced and burnished brass device in Helmholtz's laboratory in Berlin and found it wanting. The instrument-makers again strengthened the meter-length arms that served as small optical benches rigidly normal to each other, and Michelson himself tried many ways to improve his sights. He was looking for a minute shift of the pattern of light and dark bands across his eyepiece as the device was rotated through 90 degrees at the appropriate astronomically calculated time periods for his location. The theory was beautiful; the practice, disappointing.

Although significant interference fringe shifts could not be detected, Michelson was delighted that the instrument itself showed extreme precision and sensitivity. Vibrations from traffic on Berlin streets after midnight were noticeable in the interferometer's eyepiece. So Michelson took his instrument to the Astrophysical Observatory at Potsdam, and there, in the well of the equatorial of the big telescope, he finally found enough peace and quiet to get the readings he was seeking. But nothing spectacular happened, and once his data were reduced, Michelson was chagrined to find he had null results. He wrote to Bell on 17 April 1881, saying in part: "Thus the question is solved in the negative, showing that the ether in the vicinity of the earth is moving with the earth; a result in direct variance with the generally received theory of aberration."[12] Presumably Helmholtz felt that Michelson's aether-drift test was no more decisive than Rowland's electrified disk had been for the nature of electromagnetism. The only indubitable results of these two experiments were that some good talent was showing from America and that Maxwell's ideas were bearing fruit.

Heinrich Rudolf Hertz, born on 22 February 1857 in Hamburg was the first son of a rising young lawyer. The boy attended the local gymnasium, was a model student, and showed talent in architectural, mechanical, and mathematical subjects. In 1877, he went to Munich to try for a career in engineering, and there he studied with J. F. W. von Bezold. By 1878, when he went to Berlin to study at the university physical laboratory under Helmholtz and Kirchhoff, young Hertz had already turned toward pure science. He was quickly recognized by Helmholtz as a rare student, adept and insightful. When the master assigned a problem to define an upper limit to the magnitude of electric inertia in coils of wire, Hertz, as expected, took the class prize for his solution. His senior thesis on "Kinetic Energy of Electricity in Motion" was published in 1880, the same year he obtained his doctorate *summa cum laude* with another experimental study of "Induction in Rotating Spheres."

Hertz proceeded with Helmholtz's guidance to try to settle by experiments the questions raised by Faraday, Maxwell, Weber, and most German and French theorists. The immensely complicated problems of electromagnetism and electrodynamics were wrapped in conflicts between ideas of continuity and discreteness and in the various analogies used to explain the behavior of electric charges

and currents. For three years Hertz remained in Berlin as Helmholtz's assistant, trying to sort out facts from fictions in these
conflicting theories. The Berlin Academy of Sciences offered a prize
in 1879 for a solution to these conflicts between electromagnetic
forces and the dielectric polarization of insulators.[13] Hertz wrestled
with theory and experiments on contacts of elastic solids, hardness,
evaporation, and electric discharges in various rarefied gases. But
the 1879 prize eluded him (and everyone else as well) for the time
being. Growing weary of theory, he moved more and more toward
experimental physics.

In 1881, Hertz was a graduate assistant in Helmholtz's physical
laboratory at the University of Berlin when Michelson came to
work briefly there. Whether Michelson and Hertz met during this
period is not certain, but in 1883 both young men took permanent
positions, Michelson returning to the United States and the new
Case School of Applied Science in Cleveland, Ohio, and Hertz
going on to become privatdocent at the University of Kiel.

Two years later Hertz was called to Karlsruhe with a promotion
to ordinary professor of physics at the local Polytechnic. Upon arrival there in 1885, Hertz began to concentrate on studies of spark
gaps, sparks, coils, and induced sparks in secondary coils. The
climaxes of both Michelson's and Hertz's stories are reserved for
the end of this chapter. Hertz credited Bezold, Oliver Lodge, and
George Francis FitzGerald for their separate and influential studies
of electric discharges and disturbances several years previously. He
built his own oscillator and resonator in order to conduct a series of
experiments on electrostatic sparks and electromagnetic induction
currents in primary and secondary coils. As we shall see later,
Hertz fulfilled the speculations of Faraday and the predictions of
Maxwell's field ideas, simultaneously corroborating their an-
tipathies to action-at-a-distance and their affinities for some view of
space as filled with a medium or fields of force through which electrodynamic effects could be measured in time and space. For all
this, he was quickly recognized and rewarded by being called to
the University of Bonn in 1889 to succeed R. J. E. Clausius there.

During all these years the influence of Austria's Ernst Mach
and his allies was growing stronger. While Mach held two chairs at

Prague through the 1880s, his work in experimental optics, acoustics, gas dynamics, and shock waves laid a basis for respecting his authority in psychology and the philosophy of science. Mach had first taught mathematics at Graz for three years before gaining a chair of physics at Prague in 1867. Until 1895, when he returned in triumph (after a personal tragedy) to Vienna, Mach produced a wide variety of experiments, studies, and books in various disciplines and translations that gained for him a worldwide influence. He was noted above all as a leader in the effort to purge the sciences of metaphysical residues. In his 1883 study of mechanics in terms of a historical and critical analysis of its evolution, Mach argued that Newton and the Newtonians had overemphasized their absolute assumptions and so had failed to recognize that only relative motion is physically meaningful.

Mach's sensationalist psychology fitted his urge to declare meaningless all conceptions that are not based directly on perceptions. Refusing to sanction any one scientific method, Mach preached a kind of Darwinist natural selection for scientific ideas. The aim of science is to achieve "economy of thought," and positive knowledge can come only from sensible phenomena that are subject to measurement. All abstractions and generalizations are suspect unless reducible to analyses of sensations. Mach was an independent and incorrigible skeptic, although in personal relations he was kind, lovable, and humane.

Stressing observable experiences so strongly, Mach was a severe critic not only of the Newtonian absolutes of space and time but also of molecular and atomic conceptions of matter, if thought of as more substantial than heuristic models. Curiously, his anti-atomism for physics was so much stronger than his anti-aetherism that the latter rejection was so mild as to seem by comparison an acceptance. Yet Mach ultimately was so singular a philosopher and powerful a personality, even after becoming half paralyzed in 1898, that he cannot be characterized in any simple manner. Despite his later denials, he was definitely a forerunner for most of the main trends in physics and philosophy of the next century. Although his own analyses of radiation in the 1880s were limited, his iconoclastic attitudes and essays fed directly into those of the younger generation in the physical sciences all over Europe.[14]

While students of Helmholtz and Mach were at the vanguard of studies of terrestrial radiation, other students elsewhere were analyzing celestial radiations with new instruments and techniques. Inevitably, it seems, a new discipline was being created with the merger of astronomy and physics.

Lockyer's Hypotheses and Huggins's Astronomy

Joseph Norman Lockyer (1836–1920) and William Huggins (1824–1910) were rival pioneers of astrophysics who shall symbolize for us here the progress of the Victorian age in probing the heavens. Just as rays and waves from sparks or cathodes offered miniature approaches to understanding radiation, so also rays and waves from the sun, moon, planets, and stars offered maximal approaches to appreciating radiant energy transfer. Just as the complexity of microphysical structures was forcing superspecialization, so also the complex variety of the solar and stellar neighborhood was leading astronomers to specialize. Both Lockyer and Huggins tried to resist these tendencies but found the trends irresistible.

Huggins was a privately educated London observer who took up spectroscopy and photography in the 1860s and made them both practical, indeed indispensable, tools for all sorts of astronomy. Applying the dry-plate process for recording spectra in the 1870s, Huggins opened the ultraviolet window on the night skies and soon was obtaining "frozen visions" of cometary and nebular spectra. Moving rapidly far beyond Robert W. Bunsen (1811–99) and Gustav R. Kirchhoff (1824–87) in the arena of *stellar* spectroscopy, Huggins was nevertheless overtaken in the arena of *solar* spectroscopy by the civil servant Norman Lockyer. In 1864, Huggins thought he had discovered a new element, which he named "nebulium"; within a few years Lockyer likewise named a new element from a third sodium line in the solar spectrum "helium." Eventually Huggins's element was proved spurious, but Lockyer's was to be verified by the end of the century. Both men were to be knighted by Queen Victoria in 1897, the same year that Crookes was.[15]

All three well deserved this honor, but Lockyer especially had made signal contributions to a wide variety of liberal arts and sci-

Hermann von Helmholtz *(courtesy AIP Niels Bohr Library—Meggers Collection)*

Heinrich Rudolf Hertz *(courtesy AIP Niels Bohr Library)*

Albert A. Michelson *(courtesy The Smithsonian Institution)*

Hendrik A. Lorentz *(courtesy AIP Niels Bohr Library)*

ences, ever striving to lift Victoria's realm above its complacent tor-
por. His work in stellar spectroscopy, lunar photography, planetary
mapping, meteorology, and the study of sunspots, the chromo-
sphere, eclipses, stellar motions, and stellar evolution had been
paralleled by a literary career almost as prolific. He was the founding editor of *Nature*, the British weekly science news journal, and a
sparkplug in the founding of the South Kensington Science
Museum. He served as permanent secretary to the Royal Commis-
sion on Scientific Instruction and the Advancement of Science,
headed by the seventh Duke of Devonshire in the 1870s. Lockyer
headed the solar-physics and terrestrial-magnetism observations
and pushed throughout his long life for more and better educational programs in the name of national progress.

Lockyer was a controversialist, too, who advanced two speculative hypotheses that expressed his vision of where science should
be going. The first, called the "dissociation hypothesis," was proposed first in the 1870s as a system of ideas about the nature of matter; it stressed the mutability of atoms under extreme conditions of
pressure, temperature, and volume. This is sometimes regarded as
anticipating the electron theory that had evolved by the time of his
death in 1920. Lockyer's other vision, the so-called meteoritic
hypothesis, was a contribution to cosmology begun in the 1860s
and completed in 1890 with a book of that title. This work assumed
that meteorites are the major building blocks of the universe, supposedly accounting for Earth, sun and stars, galaxies, and nebulae
as well. Needless to say, it was less successful, but Lockyer's influence was nonetheless prodigious on the whole.[16]

One of Lockyer's enterprises relevant to the advent of relativity
was his promotion in the 1870s and 1880s of an illustrated handbook of popular astronomy entitled simply *The Heavens*. Written
originally by Amedée Guillemin, this work was thoroughly edited
with interpolated remarks by Lockyer. Designed for Europe's youth
and young-at-heart, the book was arranged in three parts, on the
solar system, the sidereal system, and the laws of astronomy. It
had already gone through eight editions by 1883, when a ninth was
revised by Lockyer's arch-rival Richard A. Proctor (1837–88) and
reprinted by the "Publishers in Ordinary to Her Majesty the
Queen." Balfour Stewart (1828–87) of Kew, another of five con-

tributing editors, had helped to update a previous edition, but the discussion of astronomical differences between Huggins's nebular hypothesis and Lockyer's meteoritic hypothesis was refereed by Proctor in Huggins's favor. A chapter on "Movements of Stars" showed the growth of a consensus on the problems of proper motions, the solar apex, radial velocities, and Earth's resultant vector of motion since the days of Sir William Herschel. Behind Maxwell's 1879 suggestion for what would later be called aether-drift tests lay the authority of such other astronomers as F. W. D. Argelander (1789–1875) and Otto Struve (1819–1905), who, since F. W. Bessel's (1784–1864) discovery of a true stellar parallax in 1838, had redrawn the heavens in terms of the relative motions of "fixed" stars. In 1880, when A. A. Michelson first designed and tried his interferometer, neither aberration nor the wave theory and its aether were primarily at issue. Rather, the main issues of interest were the cosmological questions: Toward what spot in the sky, and how fast, are we on this Earth around this sun traveling?[17]

The Astronomer Royal, Sir George Airy (1801–92), had recently devised a scheme of least sums for what he hoped was a hypothesis-free method of calculating the most probable elements of the sun's motion of translation. But Proctor and his associates were very critical of this, for the problem was quite likely one of the most complex in all of nature. Camille Flammarion (1842–1925) had just issued in Paris in 1880 the first edition of *Astronomie Populaire*, which began with the component motions making up Earth's resultant vector in the heavens and ended with the problem of the proper motions of stars. Ten identified components of Earth's movements had already been specified by 1880, and two more, numbers 11 and 12, below, were suspected.[18]

Twelve Identified Motions of Earth in the Heavens

1. Rotation on axis
2. Revolution around sun
3. Monthly inequality (Earth-moon nutation)
4. Precession of equinoxes
5. Nutation on axis
6. Variation in obliquity of ecliptic

7. Variation in eccentricity of orbit
8. Secular variation of perihelion
9. Perturbations from other planets
10. Barycentric inequality (sun–solar system nutation)
11. Helical translation of whole solar system
12. Internal Earth shifts and upsets

These component motions of Earth through the heavens recall Lambert's hypotheses regarding the problem of determining the ultimate cycloid of Earth's absolute motion. Although only item 11 expresses directly the issue of the "solar apex" (the direction of the sun's vector against the celestial sphere), much of nineteenth-century positional astronomy was aimed at this problem. Lockyer and Huggins, in their very different ways, were concerned, as were Airy, Flammarion, and others, with sorting out the evidence from stellar aberration, parallax determinations, and spectroscopic studies of radiation to see if analyses would lead toward synthesis of a new method for determining the absolute among all the relative motions of Earth. Thus the appeal of Michelson's first experiment for this purpose was guaranteed, even while Maxwell was suggesting some such possibility.

Kelvin's Elasticity and Weber's Electricity

By the early 1880s, Kelvin (who was then still Sir William Thomson) was the leading exponent of the hope that an elastic solid aether together with vortex atoms would eventually explain energy and matter. On the other hand, Wilhelm Eduard Weber (1804–91) was the venerable leader of the Newtonian tradition, which lived on through the similarities of gravitation to electromagnetism to dominate the traditional Continental approach toward electricity and magnetism. Between the leaders of these two schools of thought—which Josiah Willard Gibbs of Yale was soon to characterize as "the Elastic and the Electrical Theories of Light"—stood Hermann von Helmholtz, watching and waiting, thinking and testing.

Though more friendly with Kelvin than with Professor Weber of

Göttingen, Helmholtz, as we have seen, was developing his own ideas. If the Weberian tradition of most of his countrymen was positively antagonistic to what little they knew of Maxwell's ideas, Helmholtz knew also from his intimate friendship with Kelvin that the latter was neither wholeheartedly Maxwellian nor an unquestioned spokesman for some monolithic opinion of British science. In short, no single electromagnetic theory could command universal respect. The empirical data were accumulating too fast to fit, and each major theory had its faults.

Maxwell's *Treatise* (1873) was too critical of the simple and obvious notion of instantaneous action-at-a-distance, which had served so long and well as the basis for Newtonian gravitational theory. Point forces, the inverse-square law, and plane geometry using straight-line vectors as converted by Legendre, Lagrange, Laplace, and Poisson into the calculus and field-theoretical language of ordinary differential equations plus force laws—these components of the Weberian view had served beautifully since the early 1870s, at least for handling many advances in electricity and magnetism. Maxwell's respect for Faraday, his efforts to convert the luminiferous aether into an electromagnetic aether, and above all his peculiar use of the notion of a "displacement current" and a "vector potential" seemed to most German and French students of the subject simply misguided. Insofar as Weber had unified Coulomb's and Ampère's ideas around 1848, Maxwell's call for a more thorough field theory would take a long time to answer, and that seemed unnecessary, to say the least. Helmholtz had wrestled with these three contending theories since coming to Berlin in 1871. By 1879, having modified his earlier viewpoint of a potential theory without a dielectric aether, Helmholtz set up a prize competition among his students for some experimental tests to decide between the Amperean action-at-a-distance theory and the Faraday-Maxwellian theory of an electromagnetic aether.

The Berlin Academy prize (see p. 64) was a challenge to produce evidence of a relationship between electromagnetic forces and the dielectric polarization of insulators. Young Heinrich Hertz, then Helmholtz's most promising student, did not respond to this challenge immediately, partly because of personal circumstances of acquiring a degree and getting a job and partly because the real gist of Maxwell's synthesis—that *light is an electromagnetic disturbance* (and vice versa)—seemed not yet recognized. If it were true that Maxwell's whole orientation toward electromagnetism was conditioned by an optical rather than an electrical bias, then perhaps the best approach to Helmholtz's challenge would be to search Maxwell's equations for optical analogues that might apply to life-size invisible waves propagating through space. He would have to mull that over awhile.[19]

Meanwhile, Kelvin in Glasgow and Peter Guthrie Tait in Edinburgh had just revised their *Treatise on Natural Philosophy*, emphasizing once again that respectable nomenclature for science now must recognize Ampère's distinctions between the terms "kinematics," for the abstract science of purely geometrical motion, and "dynamics" for the physical science of force-fed motion. Further, dynamics was subdivided into "statics," which treats of relative rest, and "kinetics," which treats of relative motion or accelerations. The hard-won intellectual clarifications that had been achieved with the kinetic-molecular theory of gases and in thermodynamics since midcentury were not to be sacrificed to mathematical formalism, however, for Kelvin and Tait warned their students that nothing could be more fatal to progress than to consider the formulae as more real than the physical facts. The primitive concept of energy, first defined in quasi-modern form by Thomas Young about 1801 along with his interference experiments and the reintroduction of the wave theory of light, had now become the central theme of Kelvin and Tait's *Natural Philosophy*. The old style of debate around *vis viva* (literally, "live force") and *vis mortua* ("dead force") was now revised and reified into the concepts of *kinetic energy* ($KE = \frac{1}{2} mv^2$) and *potential energy*, respectively.[20]

In 1884, Kelvin visited the Western Hemisphere and for three weeks during October delivered a series of lectures on molecular dynamics and the wave theory of light at Johns Hopkins University in Baltimore. This long conference, attended by Lord Rayleigh as well as twenty-one American professors of physical science from around the country, was a major event in the promotion of pure science in the United States. When Kelvin later revised and published these talks as the *Baltimore Lectures*, they achieved even greater influence. Among those attending (H. A. Rowland was host, of course) were Albert A. Michelson from the Case School of Applied Science and Edward Williams Morley (1838–1923) from Western Reserve University in Cleveland, Ohio. The chemist Morley made the most vigorous contributions among the fellows, or the "21 Coefficients," gathered around their "esteemed Molecule." Listening to the verbalized thinking about stability in vibrating shells, forks, rings, plates, and vortex atoms, the participants heard Kelvin wrestle with the long-standing difficulties of an elastic-solid aether, fluid electricity, viscous jellies, wax, and pitch.

The Kelvin lemma that heads this chapter was delivered as part of an oft quoted credo in the final lecture, number 20:

> I never satisfy myself until I can make a mechanical model of a thing. If I can make a mechanical model I can understand it. As long as I cannot make a mechanical model all the way through I cannot understand, and that is why I cannot get the electromagnetic theory. I firmly believe in an electromagnetic theory of light and that when we understand electricity and magnetism and light, we shall see them all together as part of a whole. But I want to understand light as well as I can without introducing things that we understand even less of. That is why I take plain dynamics.[21]

Many have assumed this to be an epitome of the mechanistic, materialistic world-view. But Kelvin left an impression on his American audience that was considerably more complex and humble than that quotation implies out of context. Michelson was especially pleased to have the encouragement of both Kelvin and Lord Rayleigh to try his skill at repeating Fizeau's 1859 test of the velocity of light in a moving medium (flowing water) and at improving his own 1881 test of Earth's motion against the background of the supposed luminiferous aether. Returning to Cleveland from Baltimore with Morley, they undoubtedly shared much enthusiasm and probably laid some plans.

Throughout the latter half of the nineteenth century, George Gabriel Stokes (1819–1903) held the Lucasian Professorship of Mathematics at Cambridge. He was primarily a physicist interested in optics, viscous fluids, fluorescence/phosphorescence, geodesy, convergent series, natural theology, academic administration, and the luminiferous aether. His Burnet Lectures at Aberdeen, published in 1884, were on light. With great caution he began by comparing the particle and wave theories, then waxed more lyrical as he proceeded to explode the former and expound the latter theory. Perhaps more typical of the venerable British empiricists than Kelvin, Stokes was firmly convinced of undulations primarily, and only secondarily of the need for a medium pervading all space. With Tait, Kelvin, Preston, and many others, he shared the aethereal world-view primarily for inductive reasons.[22]

Although certain philosophical writers—notably Ernst Mach, who had just published his first edition of *The Science of Mechanics* in 1882—were actively criticizing received opinions in science during the 1880s, constructive work in trying to develop the Maxwellian synthesis was also moving ahead. In 1884, John H. Poynting (1852–1914) of Birmingham published his seminal paper "On the Transfer of Energy in the Electromagnetic Field," which provided the basic formulae connecting mechanical motion with electromagnetic forces. Oliver Heaviside (1850–1925), a self-taught electrician, was translating Maxwell into forms more meaningful for engineers and amateurs. Oliver J. Lodge (1851–1940) in Liverpool was working with electromagnetic waves in, on, or around wires, hoping likewise to translate Maxwell's *Treatise* into practical value as well as provide additional understanding.[23]

Nevertheless, the luminiferous or electromagnetic aether of boundless space remained a pictorial vision for some who could calculate its contradictions. De Volson Wood (1832–97), for instance, an American engineer at the Stevens Institute, published the following conclusion about the luminiferous aether as if it conformed to the kinetic theory of gases:

We conclude, then, that a medium whose density is such that a volume of it equal to about twenty volumes of the earth would weigh one pound, and whose tension is such that the pressure on a square mile would be about one pound, and whose specific heat is such that it would require as much heat to raise the temperature of one pound of it 1°F as it would to

raise about 2,300,000,000 tons of water the same amount, will
satisfy the requirements of nature in being able to transmit a
wave of light or heat 186,300 miles per second, and transmit
133 foot-pounds of heat energy from the sun to the earth,
each second per square foot of surface normally exposed, and
also be everywhere practically non-resisting and sensibly uniform in temperature, density, and elasticity. This medium we
call the Luminiferous Aether.[24]

Although most physicists and engineers probably felt more comfortable with Kelvin and Wood's type of speculations about the
aether, those scientists who felt more of an affinity for mathematics
and astronomy were already beginning to entertain speculations
about possibilities for four-dimensional physics. And the fourth
dimension after length, breadth, and width was generally expected
to be time.

In 1884, Joseph John Thomson (1856–1940), who was no relation
to Sir William Thomson (Kelvin), succeeded Lord Rayleigh as
Cavendish Professor at Cambridge. Rayleigh's service in that post
since Maxwell's death had raised the prestige of experimental
physics in Britain to a new level. Now J. J. Thomson was gradually
to carry that tradition to new heights. His first important work had
been an application of Maxwell's theory to the motion of a charged
sphere, showing that such a moving body should gain mass proportional to its charge and the electrostatic energy. Under Rayleigh,
J. J. Thomson had won the Adams prize for an essay on vortex
rings. He was therefore in the mainstream of British physics when
in 1885 he presided over the "Report on Electrical Theories" that
the British Association for the Advancement of Science had com-
missioned to try to make some sense out of the prevailing confusion of ideas.

J. J. Thomson's "Report" divided all competing mathematical approaches into five classes, reserving for last the Maxwell-Helmholtz
theory. First, he considered the geometrical approach that had been
developed by Ampère, Grassman, Stefan, and others without serious consideration of the conservation-of-energy principle. Next he
treated the approach begun by Gauss and championed by Weber,
Riemann, and Clausius, which assumed the importance of velocity

and acceleration in treating the action of currents. Then he considered the third and fourth classes of electrical theories as those dynamical approaches championed by a father and son. Franz Ernst Neumann (1798–1895) had invented a theory of neutral potential that neglected any dielectric medium, but Helmholtz had contributed to it. His son, Carl G. Neumann (1832–1925), professor of mathematics at Leipzig, had developed a unique theory of electricity. All these theories, J. J. Thomson felt, were inadequate compared to the fifth and last approach, dynamical with full attention to dielectric action in the surrounding medium, which Maxwell had bequeathed and Helmholtz was developing.[25]

For J. J. Thomson this concentration on the electromagnetic medium was "perhaps the most important step that has ever been made in the theory of electricity." Although the practical results of the other approaches seemed equally serviceable, J. J. Thomson was already profoundly concerned with applying dynamical principles, primarily that of the conservation of energy, to material systems "including, if necessary, the ether" so that kinetic and potential forces could be treated as if they exhibited the dynamics of rigid bodies. R. T. Glazebrook's "Report on Optical Theories," which immediately followed Thomson's "Report," was equally enthusiastic for Maxwell's electromagnetic theory of light. There was, therefore, little reason to suspect that the luminiferous aether might not become the electromagnetic aether against which the multiple motions of Earth should be measurable.

Michelson-Morley and Aether Wind

By the spring of 1885, Michelson and Morley were actively collaborating in work aimed at using the velocity of light for a test of the relative motion of Earth against the background of the luminiferous aether. Michelson had just completed a series of measurements that he was obligated to finish for Simon Newcomb, and Morley was postponing his chemical work on atomic weights of hydrogen and oxygen in order to concentrate, as he later wrote his father, on a new experiment "to see if light travels with the same velocity in all directions." But first it was imperative to recheck the trustworthiness of Fizeau's proof in the 1850s of Fresnel's two assumptions that the aether of space is stagnant and yet that moving transparent media will partially drag the universal media along with it locally.

Michelson began by writing letters to the leading authorities in his field asking their opinions on how to improve his experimental design. Josiah Willard Gibbs at Yale, E. C. Pickering at Harvard, H. A. Rowland at Johns Hopkins, and, especially over the next two years, Lord Rayleigh and Hendrik A. Lorentz (1853–1928) of Leyden were influential in sparking Michelson and Morley to return to these questions. Since his brilliant doctoral dissertation on the Maxwellian implications for reflection and refraction of light, Lorentz had become expert on conditions at the interfaces between materials, and he had set out to remedy the defects in Maxwell's theory for dispersion formulae, including its failure to explain the convection of light in moving media. J. J. Thomson, likewise, had recently contributed to the revival of interest in these questions.[26]

So Michelson and Morley plunged into this confused situation with elaborate plans to test Fresnel's predictions that light is sped onward or slowed down by traveling with or against a current of water. This required large amounts of distilled water, some complicated plumbing, and various current meters as well as a complex of precision optical equipment. Michelson suffered a "nervous breakdown" while worrying about all this in September 1885; it looked as if Morley would have to carry on alone, but Michelson recovered fully and returned to work by Christmas. By March 1886, the two had completed sixty-five trials of their experiment with many variations of the lengths of their tubes and the speeds of rushing fluids. Finding that the change in the observed velocity of light was, as Fresnel had predicted and Fizeau had first confirmed, precisely proportional to water speed, Michelson was pleased to report by letters to Kelvin and Gibbs that the aether-drag coefficient for flowing water fully confirmed to their entire satisfaction the work of Fresnel and Fizeau. The aether-drag hypothesis having been verified, the aether-drift hypothesis should surely prove verifiable also.

But, meanwhile, H. A. Lorentz had thoroughly researched the theoretical implications of Michelson's 1881 Potsdam experiment, trying to reconcile Stokes's theory with Maxwell's and with Michelson's evidence. The 1881 experiment had not attracted much attention, and indeed it had been erroneous in a few respects. Michelson received a letter of encouragement from Rayleigh in March 1887, after preparations were well along for a renewed and revised aether-drift experiment. He wrote in reply concerning his original test for aether drift in the "cellar" at Potsdam, "I have never been fully satisfied with the results of my Potsdam experiment even taking into account the correction which Lorentz points out. All that may be properly concluded from it is that (supposing the ether were really stationary) the motion of the earth thro[ugh] space cannot be very much greater than its velocity in orbit." He had not yet seen Lorentz's critique of 1886, which also urged a more elaborate and careful reptition, but he assured Rayleigh that his letter had "once more fired my enthusiasm and it has decided me to begin the work at once." Michelson promised to make further improvements "doubling or tripling the number of reflections so that the displacement would be at least half a fringe."[27]

Expanding upon Michelson's original design, the two collaborators built a much larger apparatus based on a brick pier laid on bedrock beneath a basement laboratory. In an annular trough filled with mercury, they fitted a circular float, on top of which they placed a massive sandstone slab 1.5 meters square and 0.3 meter thick. Using this slab as an optical bench they placed at each corner four plane parallel speculum-metal mirrors. Virtually monochromatic sodium light from an Argand burner was used as the source, passing through two slits and a lens to provide the "point source" beam that was aimed toward a half-silvered glass called the beam-splitter. The beam of light was thus divided equally into two perpendicular pencils of light, and these pencils, or half beams, traveled back and forth diagonally by multiple reflections over equal path lengths before being recombined at the beam-splitter for the last lap into the small telescopic eyepiece. There the two coherent pencils of light formed an interference pattern of white fringes only when the two optical paths traversed were exactly equal, which condition was controlled by a micrometer screw on one adjustable mirror beside the viewer's telescope.

Their preliminary apparatus had been almost complete when on the night of 27 October 1886 a disastrous fire struck the Case main building. Some students from a nearby dormitory of Western Reserve rescued some of the equipment for this experiment, but most of Michelson's apparatus, like the building itself, was completely destroyed. Thus Morley and Western Reserve arranged for Michelson's laboratory to be salvaged and transferred to the southeast-corner basement room of Adelbert Hall. There slowly and temporarily Case's physics department was re-established, and there finally, in July 1887, Michelson and Morley were ready to make their observations.

The heavy stone slab was set on a pier so that the optical plane was at eye level. Once set in motion it would rotate very slowly (about once every six minutes) for hours. The whole optical portion of the apparatus was covered with a wooden box to minimize temperature fluctuations from air currents. At sixteen points around the compass, the observer would reset the crosshairs in his field of view on the clearest central interference fringe by a twist of the micrometer knob at the time of passing each azimuth mark. Readings in terms of divisions on the screwheads were recorded, each division corresponding to .02 wave length, each fringe averaging about fifty divisions. Astronomical estimates of the best time to observe were unnecessary (unlike the test of April 1881) because only the orbital velocity of Earth was being sought. The path length then had been 120 cm, and now the total light paths stretched over 1,100 cm, which corresponded to 2×10^7 wave lengths of the yellow sodium light used as a gauge. Had anything been overlooked?

On Friday and Saturday, 8 and 9 July, then again on Monday and Tuesday, 11 and 12 July 1887, at noon in a counterclockwise direction and at 6 P.M. in a clockwise direction, Michelson and Morley recorded data from six turns of their apparatus. Each session lasted only an hour or so, with thirty-six minutes of viewing time the standard. Observation times were chosen to keep disturbances at a minimum while the students upstairs were out to lunch and supper.

Assuming an orbital velocity for Earth of about 30 km per second (18.6 mps) and having now measured the velocity of light at approximately 300,000 km per second (186,000 mps), the ratio between the velocity of Earth in orbit and the velocity of light was about 1/10,000. But because of the necessity of returning the light beam over its own outgoing path in all such experiments so far, this was a second-order experiment, which meant that both numerator and denominator were squared, so that the measurement ratio was actually 1/100,000,000. In 1881, Michelson had hoped to see fringe shifts of .04, whereas in 1887 he had expected a deviation from zero of .4 of a fringe. When their tabular data were reduced and redrawn in graphic form, both observers were appalled to find their maximum displacement was only .02 and the average much less than .01. Michelson and Morley ended their 1887 paper saying:

> It seems fair to conclude from the figure that if there is any displacement due to the relative motion of the earth and the luminiferous aether, this cannot be much greater than 0.01 of the distance between the fringes. . . . The actual displacement was certainly less than the twentieth part of this, and probably less than the fortieth part. But since the displacement is proportional to the square of the velocity, the relative velocity of the earth and the ether is probably less than one-sixth the earth's orbital velocity, and certainly less than one-fourth.
>
> In what precedes, only the orbital motion of the earth is considered. If this is combined with the motion of the solar system, concerning which but little is known with certainty, the result would have to be modified; and it is just possible that resultant velocity at the time of the observations was small though the chances are much against it. The experiment will therefore be repeated at intervals of three months, and thus all uncertainty will be avoided.[28]

Despite their promise to repeat the experiment at three-month intervals for all four seasons of the year, Michelson and Morley never again did this experiment together. They were both extraordinarily disappointed with their results, and each turned to more interesting and profitable tasks.

But there were other reasons as well. Michelson suffered a number of misfortunes that summer and fall. After a vacation back East to his wife Margaret's family, the Heminways, in late summer, the Michelsons returned to Cleveland in September only to find their house a shambles. Their trusted servants had absconded with most of their valuables. After the police finally apprehended the culprits, a new set of servants attempted to blackmail the handsome young professor on a charge of seducing a pretty Irish maid. When countercharges of extortion were filed, the Cleveland newspapers caught a juicy story and headlined the scandal. To the Morleys and others nearby, it was no secret that the Michelsons had been having domestic difficulties since the fall of 1885, at least, when Margaret had tried to have her husband committed to an asylum or sanitorium. All these problems plus the catastrophic fire at Case had seriously afflicted Michelson, yet he buried himself in his love for physics and was producing some of his best work despite these misfortunes.

In addition to the academic uncertainties and relocations after the Case fire in 1886 and again in the autumn of 1887, Michelson and Morley soon became completely absorbed in an entirely new and promising possibility. The interferometer principle could be made to do all sorts of things, and the most exciting immediate challenge was to follow up another of Maxwell's suggestions to find a way to measure the meter and to convert the length it represented into an invariable number of waves of some specific segment of the electromagnetic spectrum. This challenge both Michelson and Morley knew beforehand could not give a null result.

But meanwhile, in the fall of 1887, before their major report of the aether-drift fiasco went to press, Michelson prepared a "Supplement" of four pages that was filled with other ideas for investigating the problems of astronomical aberration and of the motion of the solar system through space. In the very last paragraph of the paper before the "Supplement," Michelson and Morley had claimed to have refuted Fresnel's explanation and, very elliptically, to have revived Stokes's theory of aberration. Lorentz's advanced work they probably neither knew directly nor could have understood had they had copies, judging by their own admissions in letters. Nevertheless, the last sentence of their first report was a direct challenge to the man who would soon prove to be the greatest physicist between Maxwell and Einstein: "If now it were legitimate to conclude from the present work that the ether is at rest with regard to the earth's surface, according to Lorentz there could not be a velocity potential, and his own theory also fails."

In 1889, Michelson was glad to accept a call to move to the new Clark University in Worcester, Massachusetts. There with more vigor than ever he attacked, among other things, the metrological problem of providing a more "natural" standard for lengths. The metric system's rather arbitrary one-ten millionth of a quadrant of Earth's circumference represented by the standard meter bar at Sèvres near Paris had been jeopardized, for instance, by the Franco-Prussian War. It could be based on a count of elemental wave lengths now that interferometric techniques could be combined with spectroscopy. Morley remained at Western Reserve and perfected his quantitative analyses for the atomic weights of the constituents of water and air. Michelson's successor at Case, however, Dayton C. Miller, was destined to worry over this lack of completion and other flaws in the aether-drift experiment and to provoke its repetition after 1900 and again after 1920.

The classic Michelson-Morley experiment of 1887 remained incomplete of its full seasonal tests until Miller completed them in 1926. Although the 1887 experiment remains a favorite example of a crucial experiment as well as a bone of contention among physicists and philosophers, neither Michelson, Morley, nor Miller thought of it as settling the issue of the existence of the luminiferous aether. If anything, Michelson's aether-drift work heightened controversy over the hypothetical light-bearing medium and sharpened the wits of those theoretical physicists who were most concerned with the ultimate nature of matter and motion, or atoms and electrons.[29]

Hertzian Radiation and Aether Waves

In 1885, when Hertz arrived in Karlsruhe, he set about systematically to develop the spark generators that the advent of alternating current could more easily support. Like most of his countrymen, he was still deeply immersed in the Weberian interpretation of electricity and magnetism. But thanks to his personal relationship with Helmholtz and to his direct study of Maxwell, Hertz probably knew better than anyone else the subtle differences in the three dominant approaches to electromagnetism—Weberian, Maxwellian, and Helmholtzian.

The conditions of that unclaimed 1879 Berlin Academy prize still haunted him, but he had learned to translate the propositions of 1879 into new idioms for 1886. Instruments and ideas were equally needful, he knew, to break the impasse. With sharpened instincts, Hertz aimed for the heart of the problem: if he could demonstrate in open air the existence of electromagnetic waves and their finite rate of propagation, most if not all of the rest of the confusion should clear itself up. Gradually he decided to try to make and manipulate radiation through the aether in the laboratory, then to apply the whole panoply of optical tests. If either the aether or electric waves existed, light should become demonstrably incorporated in the electromagnetic synthesis.

At Karlsruhe Polytechnic in the spring of 1886, Hertz first noticed the phenomenon of side-sparks induced in neighboring coils when batteries of Leyden jars or even small induction coils were discharged through a spark-gap. With his mind fully charged with Helmholtz's problems for at least seven years, Hertz was prepared to see the significance of this curiosity for experiments in, and theories of, electromagnetic action. Over the next two and a half years, he deliberately steered himself through a magnificent intellectual adventure and toward a thorough understanding of what was involved. He analyzed systematically, even when encountering some failures and dead ends, the surgings of alternating currents made visible through such spark-gaps. By the end of 1888, he had achieved a "natural end" to his researches and so published a classic summary paper "On Electric Radiation." His experimental autobiography tells his tale in eight preliminary papers published along the tortuous way. The essence of his first observations, "On Very Rapid Electric Oscillations," he learned in April 1887, had been anticipated by W. von Bezold in 1870. Later he heard of the theoretical suggestions of FitzGerald in 1882 and 1883 that aether-waves ought to emanate from variable electric currents. Hearing also of concurrent experimental work by Oliver Lodge, he had good reasons to make haste. But it was his own prepared mind that first led to experimental confirmation of induced sparks from primary to secondary open circuits several meters distant.[30]

In the spring and summer of 1887, Hertz produced sparking action-at-a-distance through air (or was it through aether or space?). Then after testing this phenomenon for resonances and nodes, Hertz felt justified in announcing the electric induction of rectilinear open circuits. Ultraviolet light (in what would become known as the "photoelectric effect") complicated his observations for a while, but, by the fall of 1887, he had discovered a variety of regular inductions of sparks in open circuits along straight lines as far apart as 14 m. Furthermore, he had proved these to be electromagnetic and not electrostatic forces, and he was convinced from his investigations that the interaction of these two forces demonstrated a "finite rate propagation of electrical actions." On 10 November 1887, he was able to report to the Berlin Academy the successful solution of its 1879 prize problem.

Curiously, the difficulty Hertz experienced in making visible the small secondary or induced spark led to a most remarkable wonder. He constructed a small black velvet–lined box to blot out extraneous light and made a small peeophole to see the induced spark. But to his amazement nothing could be seen inside when all ambient light was excluded. Tests with various light sources and red and blue filters soon led him to proofs that the visible sparks were *enhanced by ultraviolet* and *suppressed by infrared* illumination. This beautiful and important discovery was a true anomaly within all existing electromagnetic theories.

Wilhelm Hallwachs (1859–1922) was one of the first physicists to consider seriously Hertz's discovery of what was later dubbed the "photoelectric effect." In 1888 Hallwachs began to publish results of a series of tests on the influence of electric arc light on electrically charged bodies. Using polished zinc targets connected to a gold-leaf electroscope and a Siemans arc lamp as a source, he measured the discharge of positively charged plates. Variations in his apparatus confirmed Hertz's contention that ultraviolet (and not infrared) rays of light caused certain electric discharges whether visible or not. Schuster, Arrhenius, and Lenard among others later continued such investigations, but this phenomenon was not adequately explained even in 1905 when young Einstein took up the challenge by responding with the hypothesis of light-quanta.[31]

Meanwhile, Hertz returned to Berlin on 2 February 1888 to give his proof "On the Finite Velocity of Propagation of Electromagnetic

Actions." After measuring nodes and antinodes, compared along a wire and in "free space" interference, and after much tumultuous thought and many revised experimental designs, Hertz claimed that the velocity of his electric waves was greater through air than through wires. But then Henri Poincaré in Paris pointed out an error in calculation, and Hertz's own further experiments led him to see disturbing effects from metallic walls, an iron stove, and other peculiar features of his laboratory. Shadows seemed to exist *behind* conducting masses, and in front of them there seemed to be reinforcements for his spark-gap detector *as if* from reflected waves. This he felt Maxwell had predicted, but he still found it extremely difficult—"almost inadmissible"—to believe; it seemed tantamount to thinking that dielectrics or insulators made the best conducting media for these waves.

So Hertz set about simplifying the creation and *measurement* of electromagnetic waves in air, to obtain wave lengths directly from reflections and interferences, exhibiting "the propagation of induction through the air by wave-motion in a visible and almost tangible form." In a large physics lecture room (15 × 14 × 6 m) Hertz with only one or two assistants set up a sheet of zinc (4 × 2 m) against the wall, set his primary oscillator 13 m from this zinc wall, and used the same hand-held spark-gap detector (35 cm in radius) that he had been using for some months. In the darkened room cleared of all other metallic objects and obstructions, Hertz was able to see the brightening and dimming intensity of his resonating sparks as he moved the detector back and forth along the line between the oscillator and the zinc mirror: the half-wave length was more than 4 and less than 5 m.

Many other effects were produced by techniques analogous to those used in acoustics and optics, and so Hertz became convinced by March 1888 that the Faraday-Maxwell view of electromagnetic phenomena was essentially correct—only the velocity differences in wire and air seemed problematic. Hertz found a limiting case in Helmholtz's theory where "electrostatic force" was reduced to zero, thus agreeing with Maxwell's assumptions that only combined electromagnetic forces are propagated through space and that these coupled forces are propagated at the velocity of light. With the theory clarified for his purposes, all during the summer of 1888

Hertz studied waves in wires, between two wires, two plates, in tubular spaces, and in various metals and interposed insulators.

Thanks to prior and concurrent work by Oliver Heaviside, J. H. Poynting, and Oliver Lodge, Hertz was able to learn much more about his controlling assumptions rather quickly. In the autumn of 1888, while tuning resonators of smaller and smaller dimensions, Hertz noticed he did not need the big apparatus and larger rooms he was contemplating. He found he could now observe wave lengths as small as 24 cm or 12 cm half-waves. The resonance problems that had been complicating his observations of "long waves" so far could almost be eliminated by working solely with these new "short waves" in his laboratory. Furthermore, Poincaré (as opposed to Cornu) and soon V. Bjerknes of Norway (as opposed to several other German and Swiss physicists) reinforced Hertz's theoretical and experimental interpretations, so that by the end of 1888 Hertz was able to describe and publish his most classic expression of his work. "On Electric Radiation" was at once the greatest vindication of the Faraday-Maxwell tradition and perhaps the most awesome portent in all the history of science for the history of technology, communications, community, and society.[32]

In 1888, Hertz reported the fabulous results of the progressively more thorough work of the last two years at Karlsruhe Polytechnic. Having worked out better ways of constructing his induction coils and spark-gaps as oscillators, he created an electric wave generator, or in later terminology, a transmitter. And having tried many sizes and arrangements of secondary coils, hand-held spark-gaps, and conducting materials, he found several different kinds of detectors or resonators, or in modern terms, receivers. The pioneering nature of his discoveries meant that he had neither nouns nor verbs to express adequately his findings, and this handicap has plagued his interpreters and translators, for the field he opened bloomed so rapidly that the nomenclature and syntax of radio (originally "radiotelegraphy") and wireless waves through the "aether" (space or electromagnetic fields) never had a chance to become stabilized.

Hertz had solved one of the major difficulties standing between Maxwell's theory and radio by constructing a better oscillator, one producing sparks at more than ten times the frequency, and thus waves of less than one-tenth the length, of those he had originally produced. This allowed him to conduct all sorts of experiments directly analogous to optical experiments in the space of his laboratory rooms and with equipment no larger than could be easily handled by an average man. Here came the most crucial of Hertz's experiments, allowing him to announce confidently in 1888 that he had "succeeded in producing distinct rays of electric force, and in carrying out with them the elementary experiments which are commonly performed with light and radiant heat." His journey of discovery is one of the classic accounts of exploration for which there can be no substitute but to study the originals. Yet, since they are still difficult to find, the following summary of his analysis of radiation in terms of optics is attempted, based on his own accounts of studies in mid- and late 1888.

Maxwell's equations, as interpreted by Hertz, predicted that oscillating sparks produced by an alternating current surging very rapidly first into one, then the other of two separate primary conductors (say, polished metal balls separated by an air gap) ought to produce waves of energy spreading through the neighboring space at the speed of light. As each sphere discharged to the other, then recharged itself from the other, or changed polarities, there should be generated a field of electric radiation like the wave fronts generated by light sources. Hertz used a Ruhmkorff coil connected to two solid brass cylinders 3 cm in diameter and 26 cm long tipped by solid spheres slightly larger (4 cm in diameter), separated usually by about 3 mm. This oscillator or transmitter when energized would sputter with sparks averaging 1 to 2 cm in zigzag length between the spheres. As before, for a detector or receiver, Hertz used a simple hand-held loop of copper wire 1 mm thick and only 7.5 cm in diameter, which was interrupted by a minute air gap. There, one end of the wire supported a tiny brass sphere, and the other had a pointed tip that was held in close proximity by a fine screw insulated from the wire. These arrangements had sufficed in outline to deduce the need for reducing the wave length from 4 to 5 m to about 60 cm and for changing from a circular concave parabolic "mirror" to a hemicylindrical concave parabolic "mirror."

By 1888, Hertz had constructed a wooden frame backed by sheet zinc bent around the curvature and assuming a form 2 m in height, 1.2 m in breadth of aperture, and 0.7 m deep. Placing the primary oscillator in the middle of the focal axis of such a concave mirror to serve as the transmitter, then testing it for the standing waves produced by reflection from another flat zinc sheet (2 m × 2 m × ½ mm) set up against a wall, Hertz was able not only to detect minute secondary sparks at distances of 5 to 6 m and even 9 to 10 m in the direction of the optical axis of this transmitter but also to detect four nodal points, one at the wall and others at 33, 65, and 98 cm from it. Thus he got 33 cm as a closer approximation of the half-wave lengths he was producing, and he began to think in terms of an "electric ray proceeding from the concave mirror."[33]

These results suggested to him immediately the need for a second mirror to act as a receiver. It should be exactly like the primary one except for having a rectilinear secondary detector along its focal axis. After various trials of the best arrangements with sealing-wax and rubberbands holding the various electrical parts in place, Hertz and his assistants settled upon two straight wires 50 cm long, placed in line but not in contact in the middle of the focal axis, with two lead wires (covered with gutta-percha insulation) connected to the delicate new secondary spark-gap located directly behind this second mirror. The whole device was a grand inspiration, the first dipole antenna. Immediately Hertz was able to produce secondary sparks at will all around his rooms at distances of 16 m and more. But the two mirrors had to be aimed at each other, the knobs of the oscillator had to be kept well polished, and the detector's parts (knob and point) had to be kept carefully in adjustment, that is, extremely close together but not touching.

Now began Hertz's systematic exploitation of the potentials of his new apparatus. First, he would prove *rectilinear propagation* of his invisible electric waves, then *polarization* to show that they were *transverse* vibrations like light, then *reflection* characteristics to separate advancing from reflecting waves, and finally *refraction* characteristics to see how far the optical analogy might be carried.[34]

Rectilinear propagation. Using two large sheet-zinc screens (2 m × 1 m) to create "slits" of various widths, Hertz was able to prove the existence of sharp shadows by the behavior of his secondary spark-gap mirror. An assistant walking across the path of the ray, for instance, blotted out the tiny, sputtering detector spark. But "insulators do not stop the ray—it passes right through a wooden partition or door; and it is not without astonishment that one sees the sparks appear inside a closed room." Even *diffraction* phenomena were easy to produce but harder to measure accurately with such crude equipment.

Polarization. To demonstrate the obvious polarization of his electric ray, given the hemicylindrical parabolic reflectors he used, Hertz simply turned his receiver from the usual vertical onto its side, noting that the "two mirrors behave like the polarizer and analyser" of optical polarization apparatus. Not content with this, however, he ordered the construction of an octagonal frame 2 m in diameter across which were stretched 1-mm copper wires parallel to each other and 3 cm apart. This parallel-wire screen, when interposed perpendicularly between transmitter and receiver so that the wires ran perpendicular to the mirrors' focal axes, had no effect on the received spark. But turned through 90 degrees so that the wires ran parallel to the focal lines, the screen effectively stopped the ray completely! By many other variations, Hertz became assured that, so far, he had been able only to recognize the electric force. But he felt sure that the

> . . . results of experiments with slowly alternating currents leave no room for doubt that the electric oscillations are accompanied by oscillations of magnetic force which take place in the horizontal plane through the ray and are zero in the vertical plane. Hence the polarization of the ray does not so much consist in the occurrence of oscillations in the vertical plane, but rather in the fact that oscillations in the vertical plane are of an electrical nature, while those in the horizontal plane are of a magnetic nature.[35]

Reflection. From many playful and serious arrangements of his two flat zinc sheets, Hertz was able to prove that the reflections were regular, not diffuse. Through a doorway into an adjoining room, he beamed his ray at a reflecting flat set at 45 degrees; then he closed the door, and with his receiver in the far room at a 90-degree orientation relative to the transmitter, he aimed at the reflected ray. The stream of sparks was not interrupted at all; in fact, tests fully proved that the reflection was regular also, angles of incidences and reflection being equal, and Snell's law of refraction was also satisfied. But thenceforth there arose a curious anomaly. Hertz was not sure whether the ray after reflection continues to be plane-polarized. Using his octagonal wire screen again, and thinking of the analogy with tourmaline plates in optics, Hertz found a difference: "The tourmaline plate absorbs the part [of the light beam] which is not transmitted; our surface reflects it." Thus, eolotropism was confirmed by rotating the octagonal screen and inclining it at 45 degrees to the focal lines.

Refraction. For refraction, Hertz was inspired to design a large prism made of hard pitch, a material like asphalt. Cast in three identical parts because of its weight and for ease of handling, this isosceles triangular mass was 1.5 m high when stacked, 1.2 m on the sides, and had a refracting angle of 30 degrees. Left intact in the wooden casting boxes, this prism of pitch worked beautifully for geometrical checks of the last major optical property that Hertz wanted in order to prove Maxwell's predictions. He was even able to derive a refractive index, 1.69, for this impure prism of pitch.

In conclusion, Hertz wrote the following paragraph for his classic paper of 1888:

> We have applied the term rays of electric force to the phenomena which we have investigated. We may perhaps further designate them as rays of light of very great wavelength. The experiments described appear to me, at any rate, eminently adapted to remove any doubt as to the identity of light, radiant heat, and electromagnetic wave-motion. I believe that from now on we shall have greater confidence in making use of the advantages which this identity enables us to derive both in the study of optics and of electricity.[36]

As to the theoretical implications, Hertz was equally cautious or precise: "The object of these experiments was to test the fundamental hypotheses of the Faraday-Maxwell theory, and the result of the experiments is to confirm the fundamental hypotheses of the theory." But Hertz was fully aware of the hostility most German physicists felt toward Maxwell and also of the double meaning carried by Maxwell's use of the field concept: it could denote either pure mathematical relationships or a hypothetical medium, either of which could be imagined as a description of the energy transfer involved in electromagnetic phenomena. That hypothetical medium, the electromagnetic aether, was now proved to be a vast dielectric, but was the word "aether" merely synonymous with the words "space," "vacuum," and "field" in relation to electromagnetic lines of force? Like Oliver Heaviside, a fellow student of Maxwell, Hertz could not decide. So he compromised by asserting that the simplest and best answer to the question, What is Maxwell's theory? is that "Maxwell's theory is Maxwell's system of equations." He ended his famous introduction to the book *Electric Waves* with the following statement:

> If we wish to lend more color to the theory, there is nothing to prevent us from supplementing all this in aiding our powers of imagination by concrete representations of the various conceptions as to the nature of electric polarization, the electric current, etc. But scientific accuracy requires of us that we should in no wise confuse the simple and homely figure, as it is presented to us by nature, with the gay garment which we use to clothe it. Of our own free will we can make no change whatever in the form of the one, but the cut and color of the other we can choose as we please.[37]

Hertz's final paper in that collection was entitled "On the Fundamental Equations of Electromagnetics for Bodies in Motion" (published in 1890 in the same journal that was fifteen years later to publish Einstein's papers), to which later he added a final note, saying, "But in connection with these matters we have scarcely any right to speak of probability,—so complete is our ignorance as to possible motions of the aether."

After Hertz accepted the call to the University of Bonn as Clausius's successor, he became even more concerned with the theoretical implications of his experimental researches. Chief among those implications was the nature of the electromagnetic aether and the question of whether the fundamental idea of "force" in Newtonian physics could be discarded, particularly in regard to electrified bodies in relative motion. Hertz seems to have preferred a wager on the "hidden mass" of aether as a universal function of electromagnetic radiation, but he was extremely circumspect in how he placed this wager. As we shall see in the next chapter, Hertz was unfortunate with his primary intellectual heir, Philipp Lenard, but his experiments alone were brilliant enough to win him enduring honor and fame.

Hertz's *Principles of Mechanics* proved him to be one of the most profound of modern thinkers. Yet because he died so young and because his analysis was so foreign to most people's experience, his message was lost on most of the modern world. Not only did he see his way through the confusion of attempts at explaining electromagnetism by aiming finally to demonstrate that optical analogs existed for all these new invisible electric waves, but also he did his best to purge modern science and philosophy of naïve acceptance of the doctrine of *force* transformed into *energy* as the primary explanatory principles for nature. Hertz's *Principles* reduced to three concepts—space, time, mass—those primitive ideas necessary for an electromagnetic world-view. In this way, he hoped to link Maxwell and Helmholtz with a future philosophy of physical science that would redefine traditional mechanics. His last work is a magnificent *detour de force*. But it never really caught on, largely because so few could appreciate the need he felt for fewer primitive concepts and more sophisticated analyses of radiation and energy transfer.[38]

Success and failure in equal measure crowned the 1880s. Both Hertz's experimental demonstrations of aether-waves and Michelson-Morley's failure to find an aether-wind were recorded in 1887. This chronological coincidence goes far to explain the history of physics over the next several decades. The two experiments posed a profound conundrum to the few who could appreciate them. George Francis FitzGerald of Dublin, Oliver J. Lodge of Liverpool, Lord Kelvin of Glasgow, Lord Rayleigh of Terling, Henri Poincaré in Paris, and H. A. Lorentz in Leyden were some of those who independently pondered with profound interest the meaning of Michelson's failure and Hertz's success.

It was FitzGerald who in print first suggested (in 1889) the material-contraction hypothesis as a qualitative explanation of the fact that light seems unaffected by the direction of its travel through space and through the aether that must fill it. Oliver Lodge, a British physicist-electrician, helped bring about the application of Hertz's experiments in radiotelephony. If the connection seemed tenuous, the aether of space certainly did not, once the development of wireless telecommunications systems began. After 1890, FitzGerald and Lodge were prime exponents in Britain of the need to use the aether-waves and to eliminate the lack of aether-drift from the Michelson-Morley experiment.

NOTES

1. For guides to the intellectual background, see John Theodore Merz, *A History of European Thought in the Nineteenth Century*, 4 vols. (Edinburgh: Blackwood, 1898–1914); Charles C. Gillispie, *The Edge of Objectivity: An Essay in the History of Scientific Ideas* (Princeton: Princeton Univ. Press, 1960); Wm. P. D. Wightman, *The Growth of Scientific Ideas* (Edinburgh: Oliver & Boyd, 1950). For the role of chemistry, see Aaron J. Ihde, *The Development of Modern Chemistry* (New York: Harper & Row, 1964).
2. For more specific background to the physics of the period, see Florian Cajori, *A History of Physics in Its Elementary Branches Including the Evolution of Physical Laboratories* (New York: Macmillan, 1929); William Wilson, *A Hundred Years of Physics* (London: Duckworth, 1950); William McGucken, *Ninetheenth-Century Spectroscopy: Development of the Understanding of Spectra, 1802–1897* (Baltimore: Johns Hopkins Press, 1969); Harold I. Sharlin, *The Convergent Century: The Unification of Science in the Nineteenth Century* (New York: Abelard-Schuman, 1966) and *The Making of the Electrical Age: From the Telegraph to Automation* (London: Abelard-Schuman, 1963).
3. A. E. Woodruff, "William Crookes and the Radiometer," *ISIS* 57 (Summer 1966): 188–98; S. G. Brush and C. W. F. Everitt, "Maxwell, Osborne Reynolds, and the Radiometer," in *Historical Studies in the Physical Sciences*, vol. I, Russell McCormach, ed. (Philadelphia: Univ. of Pennsylvania Press, 1969), pp. 105–25.
4. For more details, see Stephen G. Brush, ed., *Kinetic Theory*, 2 vols. (Oxford: Pergamon, 1965–66), and his seminal article "Science and Culture in the Nineteenth Century: Thermodynamics and History," *The Graduate Journal* [University of Texas] 7 (Spring 1967): 477-565.
5. Crookes's Bakerian Lecture "On the Illumination of Lines of Electrical Pressure, and the Trajectory of Molecules" is extracted from *Philosophical Transactions, Royal Society*, Part I, p. 135 (1879), in William Francis Magie, ed., *A Source Book in Physics* (Cambridge: Harvard Univ. Press, 1935), pp. 564–76, quoted paragraphs, pp. 575–76.

6. Stoney's "electrine" idea was one of three derived units (the other two were "velocitine" and "forcine") from primary units of measurement ("lengthine," "massine," and "timine") that he had suggested in 1874 at the British Association for the Advancement of Science meeting in Belfast in order to simplify electrostatic/electromagnetic calculations and to bring "the chemical bond, which seems to be the unit of concrete Nature, . . . into its proper relation to physics." G. Johnstone Stoney, "On the Physical Units of Nature," *Philosophical Magazine,* 5th series, 11 (May 1881): 381–90. See also Charles Susskind, "Observations of Electromagnetic Wave Radiation Before Hertz," *ISIS* 55 (March 1964): 32–42.

7. Cf. Yehuda Elkana, *The Discovery of the Conservation of Energy* (Cambridge, Mass.: Harvard Univ. Press, 1974), and the two *éloges* by A. W. Rücker on "Hermann von Helmholtz, 1821–1894" and by S. P. Thompson on "Lord Kelvin (William Thomson), 1824–1907" in Bessie Z. Jones, ed., *The Golden Age of Science: Thirty Portraits of the Giants of 19th Century Science by Their Contemporaries* (New York: Simon and Schuster, 1966), pp. 436–47, 466–90. See also Silvanus P. Thomson, *The Life of William Thomson, Baron Kelvin of Largs,* 2 vols. (London: Macmillan, 1910); Robert H. Silliman, "William Thomson: Smoke Rings and Nineteenth Century Atomism," *ISIS* 54 (December 1963): 461–74.

8. Russell Kahl, ed., *Selected Writings of Hermann von Helmholtz* (Middletown: Wesleyan Univ. Press, 1971), Introduction and pp. 409–36, quotations from pp. 412–13, 415.

9. Ibid., pp. 423, 435. See also A. E. Woodruff, "The Contributions of Hermann von Helmholtz to Electrodynamics," *ISIS* 59 (Fall 1968): 300–311.

10. John D. Miller, "Rowland and the Nature of Electric Currents," *ISIS* 63 (March 1972): 5–27.

11. Dorothy Michelson Livingston, *The Master of Light: A Biography of Albert A. Michelson* (New York: Scribner's, 1973). For a more technical overview, see Jean M. Bennett, D. Theodore McAllister, and Georgia M. Cabe, "Albert A. Michelson, Dean of American Optics—Life, Contributions to Science, and Influence on Modern-Day Physics," *Applied Optics* 12 (October 1973): 2253–79.

12. For further details, see Loyd S. Swenson, Jr., *The Ethereal Aether: A History of the Michelson-Morley-Miller Aether-Drift Experiments, 1880–1930* (Austin: Univ. of Texas Press, 1972).

13. On Hertz's life, see Philipp Lenard, *Great Men of Science: A History of Scientific Progress,* H. S. Hatfield, trans., from the 2nd ed. (London: G. Bell & Sons, 1954), pp. 350–70; Rollo Appleyard, *Pioneers of Electrical Communications* (Freeport, N.Y.: Books for Libraries, 1968), pp. 109–40. See also Russell McCormmach's article on Hertz in C. C. Gillispie, ed., *DSB* (see note 3 of Chap. 1, above), vol. VI, pp. 340–50.

14. Ernst Mach, *The Science of Mechanics: A Critical and Historical Account of*

Its Development, 6th English ed. as rev. from 9th German ed. (Lasalle, Ill.: Open Court, 1960). See also John T. Blackmore, *Ernst Mach: His Life, Work, and Influence* (Berkeley: Univ. of California Press, 1972).

15. See the article on William Huggins by Herbert Dingle in Williams, *DSB*, vol. VI, pp. 540–53, which also gives a comparison of Huggins and Lockyer. For another overview of the advent of astrophysics, see Charles A. Whitney, *The Discovery of Our Galaxy* (New York: Alfred A. Knopf, 1971), especially pp. 179–279.

16. See A. J. Meadows, *Science and Controversy: A Biography of Sir Norman Lockyer* (Cambridge: MIT Press, 1972), esp. pp. 135–208.

17. Amédée Guillemin, *The Heavens: An Illustrated Handbook of Popular Astronomy*, J. Norman Lockyer, ed.; Richard A. Proctor, ed. rev. 9th edn. (London: R. Bentley & Son, 1883), esp. pp. 294–98.

18. Cf. Camille Flammarion, *Astronomie Populaire* (Paris: Marpon et Flammarion, 1880), pp. 12–80, 790–803, and intermediate editions until G. C. Flammarion and André Danjon, *The Flammarion Book of Astronomy*, A. & B. Pagel, trans. (New York: Simon and Schuster, 1964), pp. 11–53. For the best of the genre, see Agnes M. Clerke, *A Popular History of Astronomy during the Nineteenth Century* (London: Adam and Charles Black, 1908), and her article "The Sun's Motion in Space" in the 1891 *Annual Report* of the Smithsonian Institution.

19. See A. E. Woodruff, "Action-at-a-Distance in Nineteenth Century Electrodynamics," *ISIS* 53 (December 1962): 439–59; Thomas K. Simpson, "Maxwell and the Direct Experimental Test of His Electromagnetic Theory," *ISIS* 57 (Winter 1966): 411–32.

20. Thomson [Kelvin] and Tait, *Treatise on Natural Philosophy*, 2 vols., reprint of 1879 edition (orig. 1867) (Cambridge: Univ. Press, 1923); see also Henry T. Bernstein, "J. Clerk Maxwell on the History of the Kinetic Theory of Gases, 1871," *ISIS* 54 (June 1963): 206–16.

21. Sir William Thomson [Kelvin], *Notes of Lectures on Molecular Dynamics and the Wave Theory of Light* . . . [Baltimore Lectures] Stenographically reported by A. S. Hathaway . . . (Baltimore: Johns Hopkins Univ., 1884), pp. 270–71.

22. George Gabriel Stokes, *On Light* (London: Macmillan, 1884); cf. S. Tolver Preston, *Physics of the Ether* (London: E. & F. N. Spon, 1875). P. G. Tait, *Light* (Edinburgh: Black, 1884); [Tait and Balfour Stewart], *The Unseen Universe* or Physical Speculations on a Future State, 2nd ed. (New York: Macmillan, 1875). David B. Wilson, "George Gabriel Stokes on Stellar Aberration and the Luminiferous Ether," *Br. J. for Hist. of Sci.* 6 (June 1972): 57–72.

23. John Henry Poynting, "On the Transfer of Energy in the Electromagnetic Field," *Philosophical Transactions, Royal Society* 175 (1884): 343–61.

24. DeVolson Wood, *The Luminiferous Aether* (New York: Van Nostrand, 1886), p. 69. For some intimations of time as another dimension in

connection with non-Euclidean geometries and their influence on speculations, see Alfred M. Bork, "The Fourth Dimension in Nineteenth-Century Physics," *ISIS* 55 (September 1964): 326–38.

25. J. J. Thomson, "Report on Electrical Theories," *Reports to the British Association for the Advancement of Science* 35 (1885): 97–155, esp. p. 123. See also J. J. Thomson, "On Some Applications of Dynamical Principles to Physical Phenomena," *Philosophical Transactions, Royal Society*, 2nd series, 176 (1885): 307–42, esp. p. 308.

26. See Russell McCormmach's entry for H. A. Lorentz in Gillispie, *DSB*, vol. VIII, pp. 487–500.

27. See Swenson, *Ethereal Aether*, pp. 88–97. The letter from Michelson to Lord Rayleigh of 6 March 1887 was first published in R. J. Strutt, *Life of John William Strutt, Third Baron Rayleigh* (London: Edward Arnold, 1924), p. 343, then reprinted in Cajori, *A History of Physics* (Madison: Univ. of Wisconsin Press, 1968), p. 199. For more details, see Dayton C. Miller, "The Ether-Drift Experiment and the Determination of the Absolute Motion of the Earth," *Reviews of Modern Physics* 5 (July 1933): 203–42; Robert S. Shankland, "The Michelson-Morley Experiment," *American Journal of Physics* 32 (January 1964): 16–35; and "The Michelson-Morley Experiment," *Scientific American* 211 (October 1964): 107–14.

28. Original papers given in facsimile as appendices in *The Ethereal Aether*, pp. 280–81. For personal details, see D. M. Livingston, *The Master of Light, op. cit.*, pp. 130–41. For a concise "biography" of the whole-life of this experiment, see Swenson, "The Michelson-Morley-Miller Experiments Before and After 1905," *Journal for the History of Astronomy* 1 (February 1970): 56–78.

29. Cf. Gerald Holton, "Einstein, Michelson, and the 'Crucial' Experiment," *ISIS* 60 (Summer 1969): 133–97; R. S. Shankland, "Michelson's Role in the Development of Relativity," *Applied Optics* 12 (October 1973): 2280–87.

30. Heinrich Hertz, *Electric Waves, Being Researches on the Propagation of Electric Action with Finite Velocity through Space*, D. E. Jones, trans., 2nd ed. (London: Macmillan, 1900); orig. in German, 1892, and in English, 1894.

31. Roger H. Stuewer, "Hertz's Discovery of the Photoelectric Effect," *Proc. XIII International Congress of the History of Science*, Moscow, August 1971, Section VI, pp. 35–43, and Stuewer, "Non-Einsteinian Theories of the Photoelectric Effect," *Minnesota Studies in the Philosophy of Science* 5. See also W. F. Magie, *A Source Book in Physics* (Cambridge: Harvard Univ. Press, 1935), pp. 578–79.

32. Hertz, *Electric Waves*, pp. 172–85.

33. For better illustrations, see Rollo Appleyard, *Pioneers of Electrical Communications, op. cit.*, pp. 109–40.

34. For the remarkable parallels to ray and wave optics that Hertz demon-

strated, compare, e.g., the preface in Henry Crew, ed., *The Wave Theory of Light: Memoirs by Huygens, Young, and Fresnel* (New York: American Book Co., 1900), pp. viii–xiii. Hertz is quoted sequentially in what follows from *Electric Waves*, pp. 554, 556.

35. Hertz, *Electric Waves*, p. 178.
36. Ibid., p. 185.
37. Ibid., p. 28. For another expert's view of the status of studies on radiant energy, see Samuel P. Langley's presidential address to the A.A.A.S. in 1888, "The History of a Doctrine," *American Journal of Science*, 3rd series, 37 (January 1889): 1–23. For an excellent recent study, see also Hugh G. J. Aitken, *Syntony and Spark: The Origins of Radio* (New York: John Wiley & Sons, 1976).
38. Heinrich Hertz, *The Principles of Mechanics Presented in a New Form*, D. E. Jones and J. T. Walley, trans., orig. German edn. 1894, English edn. 1899, reprinted (New York: Dover, 1956) with Introductory Essay by Robert S. Cohen.

Chapter III

Aether, Electrons, Atoms (c. 1890s)

What is ether? [That is] *the* question of the physical world at the present time.

—Oliver J. Lodge, 1892

There can no longer be any doubt that light waves consist of electric vibrations in the all-pervading ether, and that the latter possesses the properties of an insulator and a magnetic medium.

—Hermann von Helmholtz, 1894

May not the new [X] rays be due to longitudinal vibrations in the ether?

—W. C. Röntgen, 1895

Matter may be and likely is a structure in the aether, but certainly aether is not a structure made of matter.

—Joseph Larmor, 1898

The decade between 1890 and 1900 was an immensely fertile period for physical science as well as for electrical technology. It was a time rich in accomplishments, stable in politics, and confident of progress. It was the beginning of the end of the Victorian age. It seemed a time when some new scientist's synthesis might unify, even more broadly and profoundly than Maxwell had, the electromagnetic world-view.

A physics of the aether and another physics of matter—at the molecular, atomic, and even subatomic levels—mushroomed so rapidly, however, that there was neither enough time nor a clear mandate for the acceptance of a revolutionary new scheme of organization for all physical knowledge. Several candidate theories were put forward as new syntheses, as we shall see in this chapter, but though very influential in at least four cases, none was general enough to gain a wide consensus. In the next chapter we shall see how ongoing analyses of the phenomena of radiation and of relationships in energy transfer during the period from 1890 to 1910 became blended into Einstein's compelling new vision. That, too, took time to establish, and so we shall have to wait until the last chapter to see the full beauty of the Einsteinian synthesis, only brought to fruition during the ugliness of World War I and its aftermath.[1]

Herein we shall concentrate on the bases for the abortive semisyntheses published by Hertz, Larmor, Lorentz, and Poincaré; on the skeptics Ostwald and Mach; and on the astounding new experimental works that crowded the theorists into fast-moving trains of thought. The three concepts in the title of this chapter—aether, electrons, and atoms—were central issues at the turn of the century, but their interrelationships carried certain contradictions that seemed to require a fresh mind to resolve them.

For most privileged young students of science around the 1890s, any interest in natural philosophy would have been recognized and encouraged by their science teachers only if their mathematical and problem-solving abilities were clearly evident from their superior performance on rigid tests and periodic examinations through at least a decade of schooling. To become a "scientist" generally meant to choose to stay in school as an authoritarian teacher for the rest of one's life. Only in chemistry or engineering of some sort were there any obvious opportunities for employment outside academia. With the dawn of the electrical age, however, nights were changing into days, power was becoming flexible, and telecommunication was becoming indispensable. Physics began to compare with chemistry in usefulness as it became the basis for electrical, and soon for electronic, engineering. As experimental physics became more of an applied science, however, it became less natural and philosophical. To the son of a small-scale businessman specializing in the manufacture of electrochemical equipment and electrophysical devices, it was natural to ask what made such things work. Einstein grew up wondering at the lack of truly satisfying answers to such questions.

Einstein's Youth—To 1900

In 1889, the ten-year-old son of Hermann and Pauline Koch Einstein matriculated at the Luitpold Gymnasium in Munich. Having attended a Catholic parochial school for his elementary education, young Albert Einstein was a quiet, reserved child who nevertheless had already shown certain aptitudes for learning. The wonder that he had experienced over the toy magnetic compass given him as a kindergartener never left him. He had been keenly interested in the religious instruction given at school, trying to reconcile that with his parents' secular attitudes. Violin lessons since the age of six began to give him much pleasure, and a toy steam engine given him by his "Uncle Jacob"—Cäser Koch—soon became even more fascinating as he learned the principles by which it operated. And yet the ultimately embarrassing philosophical and religious questions that children ask their elders remained to haunt him.

In 1891, at the age of twelve, young Einstein was introduced to Euclid's *Geometry*. This introduction to mathematics was the second abiding wonder of his young life, because in his quest for certainty, Einstein now began to ponder with the ancients *why* nature seemed to be written in the language of mathematics and *how* mathematics and physics are related. A medical student who occasionally boarded with the Einstein family, Max Talmy, noticed the depth of this curiosity in the boy and presented him with some of the better popular-science books of the time. While attending the Munich gymnasium and despising the authoritarian teaching styles there, young Einstein, thanks to Talmy's interest, began to read Aaron Bernstein's (1812–84) People's Books on Natural Science. Five or six of the twenty-one little pocket books in this set, which began with topics on Earth, light, food, and so forth, and ended with the sun, life, and the spectrum, must have been especially impressive, for Einstein remembered them long afterward. Likewise, Büchner's *Force and Matter* and Spieker's self-help book on plane geometry were loans that paid off during Einstein's adolescence in growing abilities to talk with Talmy about science, religion, and philosophy. Frequently their conversations turned to Kant, Goethe, Helmholtz, Darwin, Marx, and Wagner.[2]

Einstein as a youth was also influenced by Professor Dr. Ludwig Büchner, M.D. (1824–99), an extraordinarily successful popularizer of scientific philosophy in the direct tradition of Ludwig Feuerbach. Talmy tells of giving young Einstein a copy of Büchner's famous exposition *Force and Matter*, which was first written at Tübingen in 1855. This work, then subtitled *Empirical Studies in Natural Philosophy*, passed through fifteen editions in twenty-eight years. Then at Darmstadt, in 1884, Büchner changed the subtitle to *Principles of the Natural Order of the Universe*. The author was, like T. H. Huxley in England, a confident polemicist, who had set out to cleanse German attitudes of the remnants of *Naturphilosophie*, Hegelian and Kantian idealism, and Lutheran and Catholic pietism.[3]

Arguing against every form of supernatural explanation, Büchner gathered quotations from hundreds of men of science to reinforce his doctrine of emancipation for a self-contained humanity. Ranging broadly over disciplines as disparate as astronomy and anthropology, history and histology, theology and metaphysics, Büchner offered a potent antidote to all forms of transcendentalism. For German-speaking peoples, he popularized a relativity of knowledge that stressed neo-positivistic attitudes toward truth and a monistic confidence in man's ability to know the full truth about nature, both physical and human.

Büchner inveighed against the dogmas of militant agnostics as well as superstitious moralists, for he equated the ultimate tendencies of the two. Atheism, in the sense of freedom from the social tyranny of belief in a *personal* God, was for Büchner, following Schopenhauer and Feuerbach, the only way out of the nonsense of inherited beliefs from the Judeo-Christian heritage. Predicting the ultimate explanation of human perceptions of the complexities of nature in terms of the mechanics of molecules (physics) and of atoms (chemistry), Büchner's *Kraft und Stoff* must have seemed a liberating doctrine to many young Germans of the late nineteenth century, Max Talmy and Albert Einstein among them.

In 1894, when his family moved to Milan, young Einstein was left in Munich to finish his studies at the gymnasium, but the arrangement did not work out, and soon he crossed the Alps to join his family in Italy. He had been a recalcitrant student in high school, and after a year in Italy, he tried and failed to gain entrance to the prestigious Swiss polytechnic institute the Eidgenossische Technische Hochschule (ETH) in Zurich. That failure at the age of sixteen led him to enroll for a refresher course at the Aarau Kantonschule, where he spent two happy semesters with congenial teachers and students, especially with the Winteler family.

About this time the youthful Einstein wrote an essay that though never published, foreshadowed his growing interest in theoretical and experimental physics and in the foremost philosophical problems in the exact sciences. It was entitled simply "On the Examination of the State of the Aether in a Magnetic Field." His father wished him to become an electrical engineer, but already this essay bore evidence that the son was moving not only toward pure science but also toward purifying qualitative ideas in physical theory before attempting more quantitative experiments.[4]

In later years Einstein remembered 1895, his sixteenth year, as the time of his personal awakening through his "discovery" of the paradoxical thought-experiment regarding what should be observable if he could move along with the crest of a wave of light: "a spatially oscillatory electromagnetic field at rest." This contradiction in terms of inherited notions of mechanics remained with the young man throughout the next ten years as he struggled to understand what the leaders of physical science were thinking.

In October 1896, Einstein entered the ETH at Zurich, where he remained as a rather close to average undergraduate except in mathematics and physics until his graduation in August 1900. He enjoyed laboratory courses but disliked lectures. Depending on the notes of his friend Marcel Grossmann to keep up with his courses, Einstein preferred to read alone the original works of Mach, Kirchhoff, Helmholtz, Hertz, and Ostwald in order to try to catch up with the whole physics profession.

By the end of his sophomore year at ETH, Einstein was sure that physics, not chemistry nor mathematics, was going to be his life's passion. He wanted to construct an apparatus that could be used to measure accurately Earth's motion against the aether of space. He was unaware that others had tried to do this. His teachers, except for Adolf Hurwitz and Hermann Minkowski, were seldom able to inspire him or even to encourage his efforts to get to the vanguard.[5]

In June 1899, Einstein injured a hand in a laboratory accident and thus was forced back into his readings even more. Mach, Maxwell, Hertz, Boltzmann, and Lorentz were by now his favorite authors, insofar as they promised a deeper knowledge of nature itself. But having learned to smell out fundamental problems and propositions, Einstein, like Mach, was beginning to feel independently skeptical of the texts, fashions, and prejudices of the profession. He was thoroughly intrigued by the abstract power of thermodynamics and by the growing fecundity of electromagnetic field theory as bequeathed by Faraday and Maxwell, and as elaborated by Hertz and Lorentz. Although during his college years he was never quite sure whether he had learned enough or had in fact caught up with the most advanced thinking on these subjects, he kept pondering his own private paradox and kept wondering about energy transfer, radiation fluctuations, charged particles and the plenum, and about the electromagnetic foundations of physics.

Lodge's Views and FitzGerald's Ideas

Two of the most influential experimental physicists of the 1890s became in later life such uncompromising apologists for the aether and so despicable in their behavior for other reasons that their colleagues virtually disowned them. Oliver Joseph Lodge (1851–1940) began publishing his researches into electromagnetic wave propagation in 1875 and ran hard on the heels of Hertz throughout the 1880s in developing the implications of Maxwell's theory. Through the 1890s, Lodge became the foremost British developer of radio-telegraphy, for which he was widely honored. He became the first president of Birmingham University in 1900 and was knighted in 1902. After losing a favored son in World War I, Sir Oliver turned toward séances and, it seemed, away from science. He published much in his later years, but most of his fellows regretted the shift in his interest from physical to so-called psychical phenomena. Consequently, Lodge's role in the history of physics has been downgraded.[6]

Likewise, Philipp E. A. Lenard (1862–1947), who was Hertz's assistant at Bonn in 1892 and his posthumous editor, became the foremost investigator of photoelectricity and cathode rays shortly thereafter. He was duly honored as the eighth recipient and sole winner of the Nobel Prize in physics for 1905. An excellent experimentalist but a poor theoretician, Lenard barely missed the discovery of X rays, the proof of the atomic nucleus, and the quantification of the photoelectric effect. He considered Röntgen, Rutherford, and Einstein to have usurped credit belonging to himself. Proudly, even arrogantly, anti-Semitic, Lenard continued to do much good spectroscopy in later life, but his doctrines of "Ur-Aether" and of "Aryan physics" were adopted by Hitler's party and Lenard became the Führer of his profession in Nazi Germany, for which reason his former achievements were later denigrated or neglected.[7]

In 1889, Oliver Lodge published the first edition of his *Modern Views of Electricity*, which began with a preface saying:

The doctrine expounded in this book is the ethereal theory of electricity. Crudely one may say that as heat is a form of energy or a mode of motion, so electricity is a form of ether, or a mode of ethereal manifestation. . . . Physicists now speak intimately of ether; its reality [is] as certain as that of air. . . . What is ether? [That is] *the* question of the physical world at the present time.[8]

Through a second edition published in 1892 and a third in 1907, Lodge maintained his influential hold on the imaginations of English-reading students. From 1880 and until 1930 at least, Oliver Lodge never tired of preaching the gospel of Faraday, Maxwell, and Hertz regarding the aether and its functions: "One continuous substance filling all space: which can vibrate as light; which can be sheared into positive and negative electricity; which in whirls constitutes matter; and which transmits by continuity and not by impact every action and reaction of which matter is capable."

These were the views for which Lodge pledged his life. Although he recognized and admitted his own analogical reasoning and though he modified and extended his views to accommodate the rush of new data, he could never accept the philosophy of relativity. Einstein's formalisms and assertions that the ethereal substratum was unnecessary or superfluous were to Lodge anathema.

Early in 1891, Lodge began a series of experiments suggested by the anomalies presented by the Michelson-Morley aether-drift tests. The problem of astronomical aberration had been resurrected with a vengeance, unless some sort of ethereal viscosity could be proved to accompany moving bodies. Lodge set up two massive solid-steel flywheels on a common axis, and then arranged for a split beam of light to travel around just inside their perimeters. There should have been an observable differential in the interference fringe shifts, showing the light to have been sped on and slowed down when traveling with and against the rotation, respectively.

Even with electrified or magnetized disks, Lodge failed to find any noticeable effects on the velocity of light in these "viscosity" experiments. By May 1892, he reported to his colleagues that the null results of his own and Michelson's experiments seemed to present a serious conflict. Michelson's interpretation of the null results from his most famous experiment with Morley was based on G. G. Stokes's theory that opaque matter might carry along as it moves its own ethereal "atmosphere." But Lodge's revolving disks had seemed to argue against such a supposition. And so Lodge felt justified in claiming Hertz's work and his own as being in the mainstream of evidence for the "simple doctrine of an ether undisturbed by motion." In his view, it was Michelson's experiment that might have to be explained away.[9]

Lodge was professor of physics and mathematics at Liverpool at this time, while across the Irish Sea at Trinity College in Dublin was George Francis FitzGerald (1851–1901), a nephew of G. Johnstone Stoney's and professor of natural and experimental philosophy. FitzGerald characteristically had been first to suggest a spark-gap generator for aether-waves and first to appreciate the need for an ad hoc explanation of the Michelson-Morley experiment. A letter to the American journal *Science*, published on 17 May 1889, expressed his initial reaction:

I have read with much interest Messrs. Michelson and Morley's wonderfully delicate experiment attempting to decide the important question as to how far the ether is carried along by the earth. . . . I would suggest that almost the only hypothesis that can reconcile this opposition is that the length of material bodies changes, according as they are moving through the ether or across it, by an amount depending on the square of the ratio of their velocity to that of light.[10]

Meanwhile, there was more ferment in Dublin as well. G. Johnstone Stoney was still deeply interested in trying to find some "natural units" for electricity, no matter how small they might turn out to be. He had also turned his attention to the cause of the double lines in the spectra of gases, and in 1891 he published a lengthy paper on this subject, which built upon his previous papers of 1874 and 1881. Stoney, now firmly convinced of the existence of very small subatomic electrical charges, argued that each chemical atom of the simplest sort had at least two such charges. Having failed a decade earlier to gain currency for his word "electrines," Stoney suggested in 1891, for the first time, the word that was to become one of the most characteristic symbols of the next century:

These charges, which it will be convenient to call *electrons*, cannot be removed from the atom; but they become disguised when the atoms chemically unite. If an electron be lodged at the point . . . of a molecule which undergoes the motion described in the last chapter, the revolution of this charge will cause an electro-magnetic undulation in the surrounding aether.[11]

FitzGerald, with such ideas in mind, visited Lodge in the spring of 1892 and repeated his variable-length suggestion, which Lodge in turn duly repeated in several publications. As leading exponents of the Maxwell-Hertz doctrine both FitzGerald and Lodge were eager to see the electromagnetic approach to the aether win out over the elastic-solid approach. Kelvin (then Sir William Thomson) had recently published results mathematically deriving from Cauchy, George Green, and James MacCullagh, which reinforced his quest for a dynamical model that could be applied to a perfectly incompressible fluid or to an ideal elastic solid. In 1888, Kelvin had found a way to rule out longitudinal waves and preserve transverse waves in a "homogeneous air-less foam" of infinite dimensions; in 1889, he elaborated this quasi-rigid, or quasi-labile, theory into a "gyrostatic adynamic constitution" for his aether-foam. Its mechanics were based on a tetrahedronal skeleton framework. Kelvin's antipathy for Maxwell's electromagnetic theory was based on the desire for "plain matter-of-fact dynamics" and a "true elastic solid" or a "perfect incompressible liquid as a foundation for wave optics."

FitzGerald, however, was as skeptical of Kelvin's theory as Kelvin was of Maxwell's. Kelvin had been born in Belfast and raised in Glasgow, whereas FitzGerald had been born and raised in the "Dublin tradition," which included not only MacCullagh, Sir William Rowan Hamilton, and G. Johnstone Stoney but also some bright younger men like Thomas Preston and Fred T. Trouton. On the other hand, Kelvin's fellow Ulstermen Osborne Reynolds (1842–1912) and Joseph Larmor (1857–1942) were likewise soon to exert considerable influence on hydrodynamic and electrodynamic models for the mechanics of an aether.[12]

Kelvin at this time was rather proud about the prospects for his vortex atom theory to link physics and chemistry through the vertices of his tetrahedral arrays. But his older mentor Sir George Gabriel Stokes (1819–1903) was still the Lucasian Professor of Mathematics at Cambridge and after thirty-one years as secretary was now president of the Royal Society. Stokes's caution tempered Kelvin's enthusiasm and perhaps led directly to Lord Salisbury's famous remark a few years later that the "only function of the word *ether* has been to furnish a nominative case to the verb 'to undulate.' "

Meanwhile, an eccentric self-educated English bachelor, Oliver Heaviside (1850–1925), began corresponding with Heinrich Hertz while working out the mathematical theory for both cable and wireless telegraphy. Since at least 1885, Heaviside's translations of Maxwell's equations had helped to build bridges between theory and experiment and between science and technology. By using Newtonian forms and vector notations, Heaviside facilitated the calculations of energy and momentum transfer and reinforced the growing utility of the concept of the electromagnetic field. Having been a telegrapher, he knew the practical difficulties that remained in Maxwell's concepts of the displacement current and the vector potential. And being a strong believer in the aether as a more fundamental reality than matter, Heaviside published many papers in *The Electrician* (later collected in book form) to show that Maxwell's work had only been "the first step towards a full theory of the ether."[13]

J. J. Thomson, who had been Osborne Reynolds's student at Manchester and who in 1884 had succeeded Maxwell and Rayleigh as professor of experimental physics and head of the Cavendish Laboratory at Cambridge, felt much the same way. He too had been a firm Maxwellian since the 1870s, but his sense of scale prevented him from making facile generalizations about the electromagnetic medium or about charged corpuscles, which he was beginning to investigate in depth. In 1893, he published his influential *Notes on Recent Researches in Electricity and Magnetism*, which was billed as a sequel to Maxwell's 1873 *Treatise*. In the best British tradition, Thomson argued for the importance of models as guides to both theory and experiment, and he opted for Faraday's lines of force embedded in Maxwell's electric field as the best approach: "It is this latter view of the tubes of electrostatic induction which we shall adopt; we shall regard them as having their seat in the ether, the polarization of the particles which accompanies their passage through a dielectric being a secondary phenomenon."[14] Although the aether for J. J. Thomson seemed mostly vacuous, his search for a molecular theory of electricity using tubes of electrostatic induction was prompted largely by the need to explain phenomena where matter and the aether were involved. J. J. Thomson considered Hertz's discoveries of transcendental importance, for both pure and applied science. But his last chapter in this set of *Notes* was devoted to the asymmetries remaining in Maxwell's legacy; it was entitled simply "Electromotive Intensity in Moving Bodies." The last chapter in Hertz's book of collected papers, which was translated under the title *Electric Waves* that same year, carried a similar and portentous title: "On the Fundamental Equations of Electromagnetics for Bodies in Motion."

Also in 1893, R. T. Glazebrook in his "Review of Optical Theories," a presidential address for the British Association, bragged about the established facts of the electromagnetic theory of light and lamented the lack of knowledge about the constitution of the aether. He noted how closely Kelvin and Maxwell resembled each other mathematically and how dissimilar they were in terms of physical intuition. Like Poincaré, Glazebrook felt that he could "understand all of Maxwell except what he means by a charged body." A few years later, while preparing his little book on *Maxwell and Modern Physics*, Glazebrook, who was J. J. Thomson's assistant director of the Cavendish, listed five "destructive features" of Maxwell's original theory that had delayed its acceptance:

1) The assumption that all currents flow in closed circuits.
2) The idea of energy residing throughout the electro-magnetic fields in consequence of the strains and stresses set up in the medium by the actions to which it was subject.
3) The identification of this electromagnetic medium with the luminiferous ether, and the consequent view that light is an electro-magnetic phenomenon.
4) The view that electro-magnetic forces arise entirely from stresses and strains in the ether; the electrostatic charge of an insulated conductor being one of the forms in which the ether strain is manifested to us.
5) The notion that a dielectric under the action of electric force becomes polarized and that all electrification is the residual effect of the polarization of the dielectric.

These difficulties had hindered acceptance of Maxwell's electromagnetism at the beginning but had gradually been dispelled— at least for most English-speaking physicists. Each problem had been attacked and redefined so thoroughly in the two decades since Maxwell's *Treatise* had first appeared that Glazebrook quoted with approval Hertz's wish to restrict Maxwell's theory merely to Maxwell's system of equations. Even the notation system in Maxwell's mathematics had been revised, debated, and revised again.[15]

Another leading theoretician contemporary with FitzGerald who had matured since Maxwell and Helmholtz was Hendrik Antoon

Lorentz of Leyden. Studying the experimental work of Hertz and the theoretical lectures of Poincaré, especially *Electricité et optique*, Lorentz became converted early in 1891 from action-at-a-distance to a contiguous-action approach. His address on "Electricity and Ether" in April of that year had suggested that the world might ultimately be explained solely in terms of tiny charged particles and the aether of space. Like FitzGerald and Stoney, Lorentz began to propose aether and electrons (although he did not use this latter name for many years) as more basic concepts for physical explanations than atoms and molecules. Like J. J. Thomson and Hertz, he was concerned with the asymmetries in the electrodynamics of moving bodies. Michelson's experiments had challenged Lorentz's curiosity from the start. In 1892, he published two major papers, one laying the foundation for his theory of electrons, the other proposing a contraction hypothesis to account for man's inability to measure the relative motion of Earth against the aether. [16]

Hertz's Principles *and* Lenard's Duty

Toward the end of 1892, while H. A. Lorentz was beginning to put forward his "electron" theory and a quantitative form of the FitzGerald contraction hypothesis, Heinrich Hertz in Bonn learned that he had not long to live. Hertz was known worldwide as the man who had put the experimental capstone on the Maxwellian synthesis by discovering wireless waves in the aether of space. But few realized that Hertz had a profound urge to make a theoretical synthesis of his own before it was too late.

Hertz had just acquired a bright young assistant, Philipp E. A. Lenard, who had been studying optics and electricity under Quincke at Heidelberg. Lenard already knew a lot about cathode rays, and when Hertz suggested that a "window" in a vacuum tube might be created, Lenard grabbed the idea and soon succeeded in making discharge tubes with windows of aluminum foil, thick enough to maintain a high vacuum but thin enough to allow cathode rays to pass through into the open air, where they could be studied with greater care. Hertz and Lenard wanted to find out whether the cathode rays were primarily related to matter or to aether; now they were opening a new way to see into this question.[17]

By the beginning of 1893, Hertz had committed to writing most of his book *Principles of Mechanics*. He hoped that this treatise, by being presented in a new form, would convince physicists that the logical and philosophical foundations of their subject could be clarified, corrected, and simplified still further by a more austere and general approach to the laws of matter in motion. Following the lead of Mach and of his mentor and friend Hermann von Helmholtz (in particular his paper of 1887 "On the Physical Meaning of the Principle of Least Action") Hertz set forth as a theoretical physicist objecting to mere "mathematical exercises" to reduce dynamics (the study of force-fed motion) to kinematics (the study of abstract motion) without regard for kinetics (the mutual action between bodies in motion). Hertz and Helmholtz had long felt disturbed by the way the almost anthropomorphic concept of force had evolved into the concept of energy, and energy in turn had to be divided between kinetic and potential forms. This artificial division could be avoided, Hertz argued, if physicists would take a new axiomatic approach to the theory of rational mechanics based only on three concepts—time, space, and mass. To Hertz, as to another one of his mentors, Gustav Kirchhoff, "it almost gave pain to see atoms and their vibrations wilfully stuck in the middle of a theoretical deduction." The same could be said of the "aether" and of "electrons" at that time.[18]

In a lengthy discursive introduction, Hertz explained his program for the two severely rigorous "parts" to follow: Book I, on the *Geometry and Kinematics of Material Systems*, and Book II, on the *Mechanics of Material Systems*. He acknowledged his debt to the Kantian tradition but not to the Cartesian. His preface cited the works of J. J. Thomson, Mach, and Kelvin (William Thomson) and Tait as influential in his own thought, but curiously he denied relying on Sir William Rowan Hamilton's principle of least action for the formulation of his own fundamental law: *Systema omne liberum perseverare in statu suo quiescendi vel movendi uniformiter in directissimam* ("Every free system persists in its state of rest or of uniform motion in a straightest path").[19]

Hertz first specified three criteria by which he expected to be judged: logical *permissibility* or clarity; *correctness*, or agreement with observed motions of matter; and *appropriateness*, or simplicity with respect to properties and relations. Then he launched into a description of three complex images of the world at that time competing for allegiance: first, the traditional views of Newtonian mechanics based on notions of space, time, force, and mass. Here Hertz criticized the idea of centrifugal force out of existence, for instance, and he used Mach's critique of Newtonian metaphysics to cast doubt upon all customary conceptions.

Second, Hertz considered at length, then rejected the growing fashion to consider energy and its transformations as the new ultimate reality. Many mathematical physicists and physical chemists were so enamored by the new-found powers of the laws of thermodynamics that they were beginning to argue for a doctrine of energism or a dogma of energetics in opposition to mechanistic and materialistic world-views. Again relying on four primitive terms— space, time, mass, and energy—these advocates, such as Wilhelm Ostwald (1853–1932) and Georg Helm (1851–1923) for instance, had gone too far in Hertz's view in declaring, almost like orthodox Muslims, "Thou shalt not make unto thee any graven image, or any likeness of any thing." For Hertz, Hamilton's principle served as an exhibit of an *incorrect* application of mathematics to material systems, leading to results "which are physically false." Furthermore, the energeticists were tending to reify their concept, attributing "to energy the properties of a substance." But most emphatically Hertz objected to the perplexities introduced by dividing the notion of energy into *kinetic* aspects based on motion and *potential* aspects based on position. This was the crux of Hertz's overwhelming desire to do this work.

Third, Hertz introduced his own world-view for mechanics, which was to postulate only time, space, and mass, and thus to deprive the ideas of force and energy of fundamental status, making them instead purely derivative relations. Hertz believed that time, space, and mass were objects of experience and that considered together, these three objects obey a single fundamental law: "Every natural motion of an independent material system . . . follows with uniform velocity one of its straightest paths." This alpha-and-omega proposition allowed Hertz to argue that Hamilton's principle, which had come to exploit the geometry of space of many dimensions and thus to confuse physics, could be relegated to the realm of pure geometry. Trying to rid himself of things obscure and unintelligible, Hertz sought a world-view that would be permissible, correct, and appropriate—in short, an "intelligible image" of mechanics. This quest led him to argue for the reality of rigidity and for a plenum of concealed masses, a continuous "all-pervading medium." Carefully avoiding the word "aether," Hertz nevertheless deliberately placed his whole conceptual system in jeopardy as he invited the tests of more precise future experience to decide the issue "by tracing back the supposed action-at-a-distance to motion in an all-pervading medium whose smallest parts are subject to rigid connections. . . . This is the field in which the decisive battle between these different fundamental assumptions of mechanics must be fought out."[20]

Hertz's malignancy grew worse throughout 1893, and he died on New Year's Day 1894. Philipp Lenard undertook to edit the whole corpus of Hertz's works, and Hermann von Helmholtz wrote a special encomium and preface for Hertz's *Principles* as almost his own last act, for he died on 8 September 1894. Helmholtz, probably the most versatile and accomplished scientist-philosopher of his age, wrote lovingly of Hertz:

> There can no longer be any doubt that light-waves consist of electric vibrations in the all-pervading ether, and that the latter possesses the properties of an insulator and a magnetic medium. Electric oscillations in the ether occupy an intermediate position between the exceedingly rapid oscillations of light and the comparatively slow disturbances which are propagated by a tuning-fork when thrown into vibration; but as regards their rate of propagation, the transverse nature of their vibrations, the consequent possibility of polarizing them, their refraction and reflection, it can be shown that in all these respects they correspond completely to light and to heat rays. The electric waves only lack the power of affecting the eye, as do also the dark heat-rays, whose frequency of oscillation is not high enough for this. . . .

Amongst scientific men Heinrich Hertz has secured enduring fame by his researches. But not through his work alone
will his memory live; none of those who knew him can ever
forget his uniform modesty, his warm recognition of the
labours of others, or his genuine gratitude toward his
teachers. To him it was enough to seek after truth; and this he
did with all zeal and devotion, and without the slightest trace
of self-seeking.[21]

With the passing of both Hertz and Helmholtz, Philipp Lenard
felt a duty to edit and supervise the publication of Hertz's legacy.
Working assiduously with the English translator D. E. Jones,
Lenard managed the appearance first in German of three volumes
of Hertz's *Collected Works* (1892, 1894, 1895) and then in English
(1893, 1896, and 1899) while moving from place to place and carrying on his own experimental researches on cathode rays. In 1894,
Lenard announced his success in bringing these rays outside the
vacuum tubes in which they originated through a thin "aluminum
window."

Now for the first time, in a space of about 10 cm beside the
"Lenard window," it was possible to apply all the techniques for
investigating ultraviolet radiation directly to the questions raised by
cathode rays. Lenard quickly constructed new discharge tubes specifically for this purpose, and his phosphorescent screens and
photographs soon provided basic new data to fuel the fires of debate as to whether cathode rays were material particles or pulses
through the aether. Hertz had favored the latter view; Arthur
Schuster in England, the former. Lenard was much impressed by
Helmholtz's use of the phrase "elementary quanta of electricity,"
and he himself used the term "quanta" to speak of subatomic bits
of electricity; but by and large Lenard thought of cathode rays as
phenomena in the aether. He later regretted the scholarly duties of
editorship, which interrupted his participation in the experimental
excitement of the middle years of the 1890s.

Lenard had always been rather peripatetic. Born in Pressberg,
Hungary, in 1862, he had first studied physics at the Modern Col-
lege there under V. Klatt. Then he had moved on to Budapest,
Vienna, Berlin, and Heidelberg, where in 1886 he earned his Ph.D.

under Quincke by building a Crookes tube with a quartz window, which proved opaque to cathode rays. When he had become Hertz's assistant in 1892 at Bonn, Lenard took his new mentor's suggestions and achieved his "window" into the "furnace" that allowed the "invisible light" of cathode rays to escape into the open air. Although fascinated by the "vast new field of investigation that had opened up" before him, Lenard had time and opportunity to publish only three papers of his own during 1894 and 1895.[22]

Because he was called from Bonn to Breslau to Aachen (Aix-la-Chapelle) to Heidelberg to Kiel by 1898, Professor Lenard's most significant elaborations of the Maxwell-Hertz-Helmholtz tradition were a few specific experimental reports and the Collected Works of Hertz. Had he already been settled in a permanent position or had he been a little less conscientious in the duty he felt toward his master, Hertz, Lenard may well have turned out a different sort of work during the ferment of 1894–98, and his behavior in later life might have made for him an entirely different reputation. In any case, the raw data he provided and the new techniques he introduced during the 1890s were fundamental contributions to the physics of both matter and the aether. Recognized as such by the award of the Nobel Prize for physics in 1905, Lenard's work soon thereafter was overshadowed by Rutherford's achievements in establishing atomic and nuclear physics and by Einstein's work introducing relativistic and quantum physics as new paradigms of understanding.

Föppl, Wien, Planck, and the Energeticists

In 1894, at the same time that Lenard was publishing his comparisons of cathode rays in gases at atmospheric pressure and in "the highest vacuum" while working on their magnetic deflection and absorption properties, a number of other physicists were eagerly pursuing similar paths. In March that year, Paul Drude in Göttingen signed the preface of his treatise on the Physics of Aether and committed it to his publisher, billed as an up-to-date guide to Maxwellian-Hertzian electromagnetism. About the same time, August Föppl (1854–1924) published the first edition of a guidebook to the "Maxwellian Theory of Electricity" based on Heaviside's theoretical simplifications and Hertz's experimental corroborations. Capitalizing on the widespread fame of Hertz's work and therefore on the renewed German interest in Maxwell's synthesis, Föppl produced the basic text that was to go through many editions under different names in the next several decades. Max Abraham (1875–1922) joined Föppl to revise it in 1904; later the work became the Abraham-Becker and finally the Becker-Sauter handbook for engineers on Maxwellian electromagnetism.[23]

Drude's editorial work and expertise in optics plus Föppl's semipopular expositions of electrical engineering as a branch of technical mechanics based in physics and on experience were immensely influential. They made Maxwell acceptable in German-speaking countries. A young man from Munich deeply interested in such subjects could hardly fail to be aware that August Föppl had just been called to the Technical University of Munich. Even if the youthful Einstein in his rebellion against all things German missed Föppl as he fled to Italy, then struggled to gain entrance to the Swiss ETH, almost certainly as an undergraduate trying to teach himself physics, he must have encountered Föppl's *Maxwell*.

In the spirit of Kirchhoff, Hertz, and Mach, both the unorthodox teacher-author Föppl and the rebel student-reader Einstein strove first of all to understand the dynamics of relative motion. But in introducing Maxwell to technically but not mathematically advanced students, Föppl divided his exposition into six main sections, beginning with the vector calculus, continuing with fundamental electrical notions, and following with straight-forward accounts of ponderomotive forces, the vector potential, and energy relations in the electromagnetic field. The fifth main section of Föppl's book, however, offered a critique of relative and absolute motion that in retrospect seems extraordinarily significant. Under the title "The Electrodynamics of Moving Conductors," Föppl wrote:

The discussions of kinematics, namely of the general theory of motion, usually rest on the axiom that in the relationship of bodies to one another only relative motion is of importance. There can be no recourse to an absolute motion in space since there is absent any means to find such a motion if there is no reference object at hand from which the motion can be observed and measured. . . . According to both Maxwell's theory and the theory of optics, empty space in actuality does not exist at all. Even the so-called vacuum is filled with a medium, the ether. . . . the conception of space without this content [aether] is a contradiction, somewhat as if one tries to think of a forest without trees. The notion of completely empty space would be not at all subject to possible experience; or, in other words, we would first have to make a deep-going revision of that conception of space which has been impressed upon human thinking in its previous period of development. The decision on this question forms perhaps the most important problem of science of our time.[24]

Föppl here suggested the very same inquiry concerning the relative motions of magnets and conductors that Einstein would use eleven years later to begin his paper on the electrodynamics of moving bodies. But for the time being, neither one had thought through the possibilities of surrendering the aether or absolute motion. Yet Föppl's premonition as expressed in the last sentence quoted above must surely have affected the prescience of young Einstein wondering about what might be observable if he could travel at the speed of light.

Wilhelm Wien (1864–1928) was another student of Helmholtz's in the early 1880s who returned again and again to the most basic researches regarding radiant heat and thermodynamics. In 1886, he earned his doctorate at Berlin with a thesis on his experiments with metallic diffraction of light and on the influence of materials on the color of refracted light. After the German steel and electricity magnate Werner Siemans (1816–92) founded the Reichsanstalt, or Imperial Physico-Technical Institute, at Charlottenburg in 1884, Wien went there to work on high-temperature problems plaguing industry and on thermal radiation theory.[25]

In this setting in 1893, Wien announced his significant displacement law, which states that the wave length of maximum thermal radiation changes in a ratio that is inversely proportional to the absolute temperature of a body. This theoretical insight provided a link for determining relationships among energy, wave lengths, and temperatures of so-called ideal black bodies, that is, perfect emitters and absorbers of radiation. The next year Wien published another significant paper on temperature and the entropy of radiation, which extended the theory of thermal radiation into empty space.

Working deeply in the tradition of Ludwig Boltzmann (1844–1906), who was then moving from Munich back to Vienna, Wien complemented the advanced mathematical work of Maxwell, Boltzmann, Joseph Stefan, Willard Gibbs, Lord Rayleigh, and others who were establishing statistical mechanics. Boltzmann himself had recently turned again toward Maxwell's electromagnetic theory, publishing his own reaction to Hertz's explication in 1891. But more recently Wien's displacement law and redefinition of an ideal black body had been leading Boltzmann to reconsider the relations between electrodynamics and thermodynamics.

The Maxwellian problem of the equipartition of energy, which was so closely related to the concepts of randomness and irreversibility in the growth of the kinetic theory of gases, was revived during Boltzmann's visit to England in 1894. There debates on the physical meaning of molecular disorder had extended the notion of equipartition of energy, at G. F. FitzGerald's suggestion, to *every-thing* in the universe, including the aether. Three degrees of translational plus three of rotational freedom should apply to planets as well as particles and to atoms as well as molecules. Boltzmann later wrote from Vienna, on 28 February 1895, in a letter published in *Nature*, his authoritative answer assuming "that the whole universe is, and rests for ever, in thermal equilibrium." Central to these discussions and their sequels was the question of whether the law of entropy could or should be interpreted according to the laws of probability. Boltzmann held ever more firmly that thermodynamic phenomena were but the macroscopic reflection of atomic phenomena as regulated by both mechanical laws and the play of chance.[26]

There were many ancillary issues wrapped around these professional debates in the 1890s. One was the question of the "heat death" of the universe; another was the dispute between geologists and physicists over Earth's age; another concerned philosophical debates over continuity versus discreteness, and determinism versus chance. But perhaps most generally interesting and problematic of all such issues was the debate that was brewing over the status of the concept of *energy*.

The same factors that had prompted Hertz to present *The Principles of Mechanics* in a new context were operating elsewhere in the minds of physicists and chemists who were sensitive to the profoundest philosophical issues. Whereas Hertz had tried to reduce reliance on the concept of force by reducing dynamics to kinematics, others were trying to purify physics by getting rid of various other postulated entities such as atoms, or the aether, or reified energy. The vast new banks of scientific knowledge about the chemical elements, wireless waves, energy transfer at finite speeds through empty space, and the meaning of vacuum in the celestial and crystal spheres were a few of the new awarenesses that complicated these arguments. Ernst Mach, as both physicist and psychologist, was the archetypal puritan philosopher of science at this time, trying to purge science of both the atom and the aether as pernicious hypothetical notions.[27]

But another, younger man who had just recently (1889) been called to Berlin to occupy the chair vacated on Kirchhoff's death was more typical as a pure physicist. Max Carl Ernst Ludwig Planck (1858–1947), like Maxwell, came from a long line of cultivated gentlemen. Born in Kiel, he was educated mostly in Munich, until he discovered that "the laws of reason coincide with the laws of nature," whereupon he went to Berlin to work in physics with Helmholtz and Kirchhoff. Finding their teaching less inspiring than their writings, Planck returned to Munich for his doctoral work, finishing in 1879 with a theoretical thesis on the two most important ideas related to physical systems, namely, entropy and energy. This work criticized the great Rudolf Clausius's definition of irreversibility and led Planck into a lonely struggle for the next fifteen years trying to clarify thermodynamics. With some bitterness at the end of his long life, he remembered how dejected a theoretician could feel among experimentalists, how Boltzmann and Clausius had rejected or ignored him, and how he had lost a first in a prize competition in 1887 for siding with Helmholtz rather than with Weber in regard to the acceptance of Maxwell's electromagnetism.[28]

All through the 1880s and well into the 1890s, Planck believed that continuity must prevail over discreteness: he was openly hostile to atomism and very doubtful about probabilism. Once in Berlin, however, near the work of Wien and others at the Reichsanstalt and while editing the lectures of Kirchhoff on heat, Planck began to take Boltzmann's statistical theory of entropy more seriously. Planck had already worked long and hard on pressure-temperature-volume relations for solids, liquids, and gases, only to find that Josiah Willard Gibbs at Yale had anticipated all these state studies on the basis of physical chemistry.

In the process of these prior studies, Planck had learned to appreciate the work of F. Wilhelm Ostwald (1853–1932), a Latvian physical chemist who had come to Leipzig in 1887 with great enthusiasm, a strong literary capacity, and a deep interest in catalysis. Together with a Swede, Svante A. Arrhenius (1859–1927), who had just worked out a theory of the dissociation of molecules for electrolysis, and the Dutchman J. H. van't Hoff (1852–1911), who was likewise interested in the theory of solutions, stereochemistry, and osmotic pressure, Ostwald founded a new journal for physical chemistry in 1887. He was also instrumental in making Gibbs known throughout Europe as one of the most influential scientists in the world.[29]

All this Planck could appreciate, but when Ostwald joined with Georg Helm, the physical chemist from Dresden, to launch an attack on the ideologies of mechanism and materialism in all of science, Planck began to doubt the direction of this thrust. As late as 1891, he had sided with Ostwald against Boltzmann, but when Ostwald's *Studies in Energetics* (1892) appeared to substitute for Gauss's and Hertz's time, space, and mass the primitive concepts of time, space, and *energy*, Planck began to entertain private objections to this hypostatization of energy.

The systematic substitution of energy for all notions of matter had been suggested as long ago as 1854 by W. J. M. Rankine (1820–72), a Scottish engineer and applied physicist who had also worried over the artificial notion of stored or potential energy.

Rankine had proposed a "science of energetics," and now Ostwald seemed on the verge of reducing all matter to manifestations of complex interactions of energy.

The chemists and physical chemists to whom the doctrine of energetics seemed most appealing were proud of the achievements of classical thermodynamics, especially of the liquid-state, triple-point, and phase-rule interactions in thermochemistry. Physicists and chemicophysicists, on the other hand, seemed more proud of the kinetic-molecular theory of the gaseous state, with its atomistic assumptions, because of its promise for the determination of molecular sizes and numbers. Its success in thermal-conduction and -convection studies could hardly be challenged.

Since the earliest works by Maxwell, Boltzmann, and Clausius on the interrelationships between thermodynamics and the kinetic theory, however, physicists had abandoned hope of following the individual motions of specific molecules or atoms. Rather, they had learned to operate with input-output analyses and to average over all possible states of a system, from initial to final, in order to calculate thermodynamic properties. This process, the so-called ergodic hypothesis (*ergo* = therefore), is intrinsically probabilistic, and in Boltzmann's view, it was atomistic as well. Because it seemed incompatible with a thermodynamic equilibrium state of maximum entropy, the "ergodic hypothesis" was a basic issue in the quarrel between "atomists" and "energeticists."

In September 1895 at Lübeck in northern Germany, the sixty-seventh annual meeting of the German Society of Scientists and Physicians took place. There Ostwald and Boltzmann locked horns to begin a dramatic controversy that lasted over the next decade. Ostwald spoke on "The Conquest of Scientific Materialism," and Boltzmann replied with an effort to demonstrate the conquest of scientific energeticism. Georg Helm spoke also "On the Present State of Energetics," whereas Victor Meyer of Heidelberg delivered an invited lecture on "Problems of Atomists." Superficially, the debate may have seemed like a contest over the ontological status of "matter" and "energy," but the participants were locked into a contest even more fundamental and multilayered. In part it was a professional contest between chemists and physicists; in part, a fight between differing metaphysical and epistemological views;

and in part this confrontation foreshadowed future commitments to either a predestined or a statistical world-view. Mach was not present and Planck was mute at this meeting, but within a few weeks the latter's critique appeared in print.[30]

Ever since his work on his doctoral dissertation, Planck had been convinced of the irreversibility of certain processes in nature. Thus he could not accept Boltzmann's statistical interpretation, which would permit the possibility of reversibility. Nor could he accept Ostwald's reduction of the science of mechanics to that of energetics, for this failed to recognize the basic role of the entropy function in dealing with irreversible phenomena.

Within half a decade after the Lübeck meeting, Boltzmann would go on to combine with new clarity classical thermodynamics and mechanics into statistical mechanics. Likewise, Planck was to be converted to atomism, not only for matter but also for radiation, and ultimately he too would embrace Gibbs's *Elementary Principles of Statistical Mechanics* (1902). But this was not to happen for a while yet, nor would the ensuing years be easy for either theoretician. Like the concept of an electromagnetic aether, the concept of Energy, capitalized, was captivating, fascinating, enchanting to most of those who wrestled professionally with the paradigms of their time.

Lorentz's Treatise *and* Larmor's Book

Hendrik Antoon Lorentz (1853–1928) and Joseph Larmor (1857–1942) were two theoretical physicists who, except for their personalities and certain minor circumstances, should have had more nearly equal influences on the advent of the electromagnetic world-view and subsequent acceptance of the theory of relativity. Like most other theoretical physicists in the 1890s, both men were deeply concerned with the implications of Hertz's positive demonstration of aether-waves and of Michelson's null results for an aether-wind. In the early years of the decade, both wrestled with the relationships between "charged particles" so small (and yet finite) as to be apparently subatomic in size and so swift (yet slower than the speed of light) and straight as to be evidently particles and not waves. Both thought the bending of cathode rays in magnetic fields must mean that such "rays" are particles with a determinable electrodynamic mass. And both believed that molecules of ponderable matter are almost certainly entirely too gross to be mechanically affected by the ultrafine charged particles. Electromagnetic effects, however, between atomically charged matter and the aether of space seemed to both men the primary problem for physics to solve before 1900.

Lorentz was an accomplished linguist, cosmopolitan in outlook, sociable, critical, and creative in equal measure. Since 1877, he had held one of the first chairs in Europe, at Leyden, specifically for theoretical physics. Larmor, four years junior to Lorentz, was a brilliant Scottish-Irish mathematician who had distinguished himself at St. John's College, Cambridge, gone to Queen's University, Galway, for five years, and then returned to Cambridge in 1885. Rather shy and diffident in his personal relations, Larmor remained a bachelor and eventually (in 1903) succeeded G. G. Stokes as Lucasian Professor of Mathematics. While J. J. Thomson taught experimental physics at Cambridge and was beginning to provide much of the data on electronic cathode rays, Larmor worked much like Lorentz in constructive criticism of proffered theories and in the creation of a theory of his own.[31]

In 1893, Larmor presented the British Association with a report on the action of magnetism on light. The next year he began the publication in three parts of "A Dynamical Theory of the Electric and Luminiferous Medium," which was continued in 1896 and completed in 1898. This work, elaborated into a lengthy manuscript, won the Adams Prize that year and was finally published in 1900 as a book under the title *Aether and Matter*. Larmor had conceived the electrodynamic definition of mass in 1893 and began to use Stoney's idea of "electron" in 1894. Stimulated by FitzGerald's many fertile suggestions growing out of their common heritage in the works of MacCullagh, W. R. Hamilton, Maxwell, and Kelvin, Larmor anticipated Lorentz in many ways regarding electron theory, but he had not the prestige, personality, or persistence of his rival.[32]

Lorentz admired Fresnel first, then Maxwell, Helmholtz, and Hertz as the mainstream of thinkers and doers in electromagnetic theory. By 1890, Lorentz had written two textbooks and many papers on electromagnetism, light, and molecular-kinetic theory and had worried hard over Hertz's waves and the contending theories purporting to explain electricity and magnetism. In 1891, a Danish physicist with a similar name, Ludwig Valentine Lorenz (1829–91), died in Copenhagen, leaving a legacy, which H. A. Lorentz had likewise independently accumulated; they shared a profound interest in electromagnetic and optical theory.

By 1892, Lorentz had declared himself converted to the Maxwellian-Hertzian doctrine of contiguous action through an electromagnetic aether. But he still felt certain reservations about both Maxwell's and Hertz's interpretations of the mysteries of field-and-particle dualities. Setting out to fuse the best features from both Continental and British electrodynamics, Lorentz developed a clear and complete distinction between "charged particles" (which he would call "ions" in 1895 and "electrons" only after 1899), the electromagnetic field with its seat in the stationary aether, and molecules of ponderable matter. This thorough separation, eventually to be called the dualism of electron and field, had much appeal as a logical clarification of the physics of radiant energy.

Also in 1892, Lorentz had returned to the study of astronomical aberration and of the null results of Michelson-Morley "aether-drift" tests. The so-called aether-drag tests of 1886, by which Michelson and Morley first corroborated the Fresnel prediction and the Fizeau fulfillment of a partly dragged aether, were to him as significant as the aether-drift tests.

Lorentz postulated an electromagnetic aether that moved even less than Fresnel's medium and would be completely transparent to the flow of ponderable matter through it. Despite this effort to uncouple the aether of space and the electrons from material molecules that run through it, Lorentz worried long about Michelson's experiments and hoped above all to solve eventually more satisfactorily than with the FitzGerald-Lorentz contraction hypothesis the problem of second-order aether-drift tests.

In 1895, there appeared from the press of E. J. Brill in Leyden a 140-page booklet that was destined to thrill the physics community with a series of compelling insights. This was Lorentz's soon to become famous *Versuch einer Theorie der electrischen und optischen*

Erscheinungen in bewegten Körpen (Treatise on the Theory of Electrical and Optical Phenomena in Moving Bodies). Lorentz's *Treatise* of 1895 merely postulated, rather than derived, the basic electromagnetic field equations, and it presented them in the compact form of vector notation:

$$\text{div } d = \rho$$
$$\text{div } H = O$$
$$\text{rot } H = 4\pi(\rho v + d)$$
$$-4\pi c^2 \text{ rot } d = H$$

where d is the dielectric displacement, H is the magnetic force, v is the velocity of the electric charge, ρ is the electric charge density, and c is the velocity of light. Lorentz provided a fifth equation as his own major contribution beyond the Maxwellian four above; this was the connecting link between the continuous field and the discrete particles of electricity, which Lorentz was now wont to call "ions":

$$E = 4\pi c^2 d + v \times H$$

Lorentz presented in this work a systematic survey of the aberration problem and the effect of Earth's motion through the stationary aether on measures of the velocity of light. The approximate form for the contraction hypothesis that had been put forth in 1892 was now refined exactly to a foreshortening factor of

$$\sqrt{1 - v^2/c^2}$$

This contraction of all material bodies in the direction of Earth's motion through Fresnel's and Lorentz's stationary aether seemed a necessary if not sufficient explanation for Michelson's null results.[33]

But not until 1899 did Lorentz seriously consider how to handle second-order effects. By then he was aware of rival explanations from Larmor, Emil Wiechert of Göttingen, Alfred Lienard, and Henri Poincaré. But thanks to the discovery in 1896 by his own student Pieter Zeeman (1865–1943) that a strong magnetic field widens the emission lines in the spectra of chemical elements, Lorentz and Zeeman found both confirmation and extensions for Lorentz's electron theory. Emission spectra after 1897 could often be resolved into triplets, and "Zeeman's effect," as it soon came to be called, offered a powerful new technique for studying radiation and the vibration patterns of various atoms. Lorentz was deeply impressed by the enormous size of the charge as compared to the mass of these subatomic particles, now called "electrons," and like Larmor, he introduced the idea that their mass ought to vary with their velocity. This revolutionary idea implied that all mass should vary with velocity through space. Thus the concept of mass itself appeared not so fundamental and immutable as was previously thought. Thoroughly convinced by now that cathode rays were simply streams of negatively charged particles, or electrons, Lorentz began to investigate more intensively what the study of electrons might be able to teach regarding the structure of atoms and thus of matter and the aether.

In 1902, Lorentz and Zeeman together were honored with the second award of the Nobel Prize for physics. Their researches into the "influence of magnetism upon radiation phenomena" were recognized by the Swedish Academy as profoundly important. On 11 December 1902, Lorentz delivered his acceptance lecture, which beautifully and humbly outlined the state of knowledge—or, rather, of ignorance—at that time regarding aether, electrons, and atoms. One paragraph read as follows:

> Permit me now to draw your attention to the ether. Since we learnt to consider this as the transmitter not only of optical but also of electromagnetic phenomena, the problem of its nature became more pressing than ever. Must we imagine the ether as an elastic medium of very low density, composed of atoms which are very small compared with ordinary ones? Is it perhaps an incompressible, frictionless fluid, which moves in accordance with the equations of hydrodynamics, and in which therefore there may be various turbulent motions? Or must we think of it as a kind of jelly, half liquid, half solid?[34]

Because of the null results of all second-order experiments on the relative motion of Earth and aether, Lorentz was inclined to believe that every individual molecule of matter is completely permeable to the passage of the stationary aether through it. The same notion ought to be true of each atom.

> . . . and this leads us to the idea that an atom is in the last resort some sort of local modification of the omnipresent ether, a modification which can shift from place to place without the medium itself altering its position. Having reached this point, we can consider the ether as a substance of a completely distinctive nature, completely different from all ponderable matter. With regard to its inner constitution, in the present state of our knowledge it is very difficult for us to give an adequate picture of it.

Thus was Lorentz thinking at the turn of the century when Joseph Larmor, even more radically, was trying to explain electrons as singularities in the aether and thus reduce all physical reality to the ubiquitous continuum of the electromagnetic aether. Larmor tried to incorporate all the best thought of this time into one eclectic theory to explain particles in terms of a plenum:

> Matter may be and likely is a structure in the aether, but certainly aether is not a structure made of matter. This introduction of a suprasensual aetherial medium, which is not the same as matter, may of course be described as leaving reality behind us: and so in fact may every result of thought be described which is more than a record or comparison of sensations.[35]

He too dealt with Hertz's and Michelson's experiments as central concerns, but unlike Lorentz, Larmor felt that mathematics provided a better guide to the future than physics per se: "All that is known (or perhaps need be known) of the aether itself may be formulated as a scheme of differential equations defining the properties of a *continuum* in space, which it would be gratuitous to further explain by any complications of structure."

In several important respects, Larmor anticipated the work and thought of his peers and colleagues. He seems not only to have preceded Lorentz in the electromagnetic definition of mass and in the exact formulation of the second-order transformation equations but also to have anticipated some of the work of Max Planck and others in moving toward quantum conceptions of particle-field relations. Larmor's monistic emphases on the aether, however, stood in rather stark contrast to Lorentz's dualistic separation of electrons and the aether (or the electromagnetic field). Thus, Larmor's claim (never very forcefully advocated) to historical recognition for his pioneering work in the theories of aether and electricity was largely eclipsed by Lorentz's *Treatise* of 1895 and its subsequent development, as well as by the enormous prestige of Poincaré's prolific influences.

Röntgen Rays, Radioactivity, and Radium

Wilhelm Conrad Röntgen (1845–1923), professor of physics at Würzberg, created one of the greatest scientific sensations of all time when he published in the December 1895 issue of his local journal a short notice of his discovery of a "new form of radiation." His mind had been well prepared by the teachings of Clausius, A. E. E. Kundt (1839–94), and F. W. G. Kohlrausch (1840–1910). Also having worked broadly in experimental physics since 1870 and in cathode-ray studies since at least 1890, Röntgen was primed to discover, as he did on Friday, 8 November 1895, that cathode and canal rays were not the only emanations from high-vacuum tubes. If the discharge from a light induction coil is passed through a covered cathode-ray tube, fluorescent screens coated with barium platinocyanide several meters away could be made to glow and show images of all sorts of opaque materials placed between the tube and the screen. Almost everything seemed transparent to these rays. They could cast shadows of more dense materials embedded within less dense substances, and these differential shadows could be photographed.[36]

Newsmen picked up one sentence in Röntgen's report especially for its medical implications: "If the hand be held between the discharge-tube and the screen, the darker shadow of the bones is

seen within the slightly dark shadow-image of the hand itself."
This news flashed around the world by telegraph led physicists everywhere to repeat and verify Röntgen's experiment within weeks
thereafter. Although Röntgen's own initial report followed closely
the procedures of Hertz's classic experiments and the interpretations that Lenard had given for cathode rays as phenomena in the
aether, Röntgen proved his rays were different from "ordinary"
cathode rays, not only because of their penetrative power, their
immunity to magnetic deflection, and their apparent immunity to
reflection, refraction, diffraction, and polarization but also because
they originated not at the cathode itself but "at that spot on the
wall of the discharge-tube which fluoresces the strongest." With
many fascinating photographs already in his files to prove the
point of rectilinear propagation through several meters' distance in
air, Röntgen concluded that these rays, which, "for brevity's sake,"
he had started to call "X rays," were neither cathode nor canal nor
ultraviolet-light rays:

> There seems to exist some kind of relationship between the
> new rays and light rays; at least this is indicated by the forma-
> tion of shadows, the fluorescence and chemical action pro-
> duced by them both. Now, we have known for a long time
> that there can be in the ether longitudinal vibrations; and, ac-
> cording to the views of different physicists, these vibrations
> must exist. Their existence, it is true, has not been proved up
> to the present, and consequently their properties have not
> been investigated by experiment.
>
> Ought not, therefore, the new rays to be ascribed to lon-
> gitudinal vibrations in the ether?
>
> I must confess that in the course of the investigation I have
> become more and more confident of the correctness of this
> idea, and so, therefore permit myself to announce this conjec-
> ture, although I am perfectly aware that the explanation given
> still needs further confirmation.[37]

While Röntgen continued through the next two years in a classic
series of papers to lead the way into X-ray research, Antoine Henri
Becquerel (1852–1908), from a distinguished family of French scien-

tist and physicians, the son of Edmond Becquerel (1820–91), who had been a peer of G. G. Stokes's in studies of phosphorescence and fluorescence, was led to discover and announce the phenomenon of natural radioactivity. After discussions with Poincaré regarding the significance of Röntgen's rays, Becquerel toward the end of February 1896 found that some uranium salts inherited from his father had thoroughly exposed a photographic plate in a dark drawer of a case. After a few controlled experiments with these encrusted salts exposing film and forming images through opaque paper, thin glass, and aluminum plates, Becquerel reported the phenomenon of spontaneous radioactivity.[38]

Without exposure to external light sources, uranium crystals could produce the same photographic effects as Röntgen's X rays. Furthermore, Becquerel soon showed that metallic uranium was even more effective than salts in this regard, as well as in discharging electrified bodies at some distances from the source. But, curiously, Becquerel's rays were soon shown to be different from Röntgen's because they could be deflected by electromagnetic fields. Much of the scientific (as opposed to technological) excitement in developing radiation studies over the next few years derived directly from this distinction. Röntgen's and Becquerel's rays were very similar in most but not all respects, and the same could be said for cathode and canal rays.

Pierre Curie (1859–1906) and Marya Sklodowska (1867–1934) were married on 26 July 1895; thus began one of the most famous and intellectually fertile marriages in the history of science. Pierre, together with his brother Jacques Curie, had already made a name for himself in 1880 when they had first discovered the piezoelectric effect, an electric differential produced in asymmetric crystals under the influence of pressure, and its converse, electrostriction. Marya, now Marie Curie, after a determined and romantic climb from Poland to Paris, chose for her doctoral work at the Sorbonne to follow up the chemistry of Becquerel's discoveries. Beginning in mid-1896, she systematically examined the heavy elements and mineral compounds containing uranium or thorium. Soon she discovered that certain ores, especially pitchblende, were more than twice as active in emitting Becquerel rays than uranium crystals. She named this process "radioactivity." Using Pierre's piezoelectric apparatus and electrometer to measure the ionization of air produced by radioactive samples, Madame Curie was soon impelled to postulate a new element.[39]

At this point, in 1898, Pierre decided to abandon his own research and join Marie full-time in progressive purifications of radioactive substances. By July of that year they had isolated a small pinch of powder that was about four hundred times more radioactive than uranium. This they named *"polonium,* after the name of the native country of one of us."[40] Similar to bismuth, polonium was spectacular enough, but by December they had isolated another substance, completely different in its chemistry because it is similar to almost pure barium, which, in chloride form, is some nine hundred times more radioactive than uranium. This they named *radium.*

Pierre concentrated on the physics, Marie on the chemistry of their joint researches. For the next four years, they toiled with their group-separation process to produce an ever better barium precipitate in which was concentrated a tiny but growing sample of pure radium. Eventually from eight tons of pitchblende they extracted one gram of the precious new element. Here was solid matter constantly disintegrating into energy!

Great intellectual ferment surrounded J. J. Thomson in 1897 when he produced his classic paper on "Cathode Rays," in which he argued that these rays "far from being wholly aetherial . . . are in fact wholly material. . . . they mark the paths of particles of matter charged with negative electricity." J. J. Thomson's ingenious experiments with three new types of vacuum tubes were designed, first, to verify "that something charged with negative electricity is shot off from the cathode," as Jean B. Perrin (1870–1942) had asserted on the basis of some controversial experiments in 1895. Second, Thomson designed a high-vacuum tube to show the deflection of cathode rays by an *electrostatic* (as opposed to electromagnetic) field. Third, Thomson also successfully achieved the magnetic deflection of cathode rays in different gases, by using a special bell jar placed between two large parallel coils of a type of galvanometer. Thus he was able to say:

As the cathode rays carry a charge of negative electricity, are deflected by an electrostatic force as if they were negatively electrified, and are acted on by a magnetic force in just the way in which this force would act on a negatively electrified body moving along the path of these rays, I can see no escape from the conclusion that they are charges of negative electricity carried by particles of matter. The question next arises, What are these particles? are they atoms, or molecules, or matter in a still finer state of subdivision?[41]

Using slight modifications of the same apparatus, J. J. Thomson went on to measure the ratio of the mass of these particles to the charges they carried. From two independent methods, he concluded that the value of the mass-to-charge (m/e) ratio for this something that was elsewhere beginning to be called the "electron" was about 10^{-7}.

Refining this value and these experiments over the next several years, Thomson and his prize students at the Cavendish Laboratory were able to provide further exact evidence for the corpuscular view of cathode rays or beta particles. But both X rays at one end of Maxwell's spectrum and Hertzian waves at the other remained elusive and anomalous, not yet closely enough identified to span the enormous gaps in physical properties between them.

At the end of 1899, Thomson published another paper, which showed cautious conservatism regarding the "stream of negative electrification which constitutes the cathode rays." Agreeing in quantitative measurements with Lenard and Kaufmann, Thomson now was most eager to point out that the discrete invariant charge e had been shown to be the same whether from ultraviolet light, carbon filaments, or cathode rays:

We have clear proof that the ions have a very much smaller mass than ordinary atoms; so that in the convection of negative electricity at low pressures we have something smaller even than the atom, something which involves the splitting up of the atom, inasmuch as we have taken from it a part, though only a small one, of its mass.[42]

With his sixteen science scholars in the 1898 "class" at the

Cavendish Laboratory, Thomson encouraged experimental research into properties of the aether, atoms, and molecules. Being flanked himself by Stokes and Larmor, Rutherford and Richardson, and Rayleigh and Ramsay, he never lost his metaphysical love for the aether, but he certainly was instrumental in pushing the boundary between physics and metaphysics farther away.

Ernest Rutherford (1871–1937), the bright son of a New Zealand pioneer, was one of the foremost members of Thomson's class of 1898 at the Cavendish Laboratory. He had arrived in England in 1895 with the intention of doing pure research into Hertzian waves, but when the news of Röntgen's and Becquerel's rays broke, Rutherford decided to look in breadth at the rash of scientific publications concerning X rays and radioactivity. Before going to McGill University in Montreal, Canada, in 1898, he published his third paper, which led him into his life's work as the founder of nuclear chemistry and physics.

By the turn of the century, Rutherford codified the emerging data on radioactive substances by naming their emissions with Greek letters as alpha, beta, and gamma rays, based on their behavior in electric or magnetic fields. Alpha rays are similar to Goldstein's canal rays and behave like postively charged particles about a thousand times smaller than hydrogen atoms. Beta rays are similar to, if not the same as, cathode rays and behave like negatively charged particles with a mass about two thousand times smaller than that of hydrogen atoms. Gamma rays were long the most mysterious of the three kinds of emissions.[43]

Noble Gases and the "Mystery of Mysteries"

Lord Rayleigh (John William Strutt, 1842–1919), while professor of natural philosophy at the Royal Institution in London, demonstrated before the Royal Society (19 April 1894) that chemically prepared nitrogen was always slightly less dense than the same elemental gas derived directly from the air. William Ramsay (1852–1916), professor of chemistry at University College, London, challenged Rayleigh's view that the explanation was contamination of chemical nitrogen by some lighter gas. Ramsay held that the more likely reason was the presence of a heavier gas in atmospheric nitrogen. With Rayleigh's encouragement, Ramsay had proved himself correct by the end of that summer. He isolated a residual gas by treating atmospheric nitrogen with heated magnesium to form a nitride, which sample had a density of 19 compared with 14 for airborne nitrogen, and spectroscopically the sample seemed unique. Meanwhile, Rayleigh too had done some more research on this matter to satisfy himself that the residue was neither oxygen nor nitrogen.[44]

Thus, Ramsay and Rayleigh joined forces and completed their thorough quantitative analyses in time to announce by the end of January 1895 the discovery of a new, completely inert gaseous element. This gas they called *argon* ("idle"), and for its discovery Rayleigh and Ramsay were awarded the Nobel prizes in physics and chemistry, respectively, in 1904. Meanwhile, however, Ramsay alone had gone on in 1895 to rediscover Lockyer's *helium* on Earth. By the end of 1898 Ramsay had discovered *krypton* ("hidden"), *neon* ("new"), and, in conjunction with M. W. Travers, who helped with the fractional distillation of liquefied air, *xenon* ("stranger").

These discoveries of new elements to fit "at the end of the first column of the Periodic Table" had been the objects of Ramsay's quest since at least 1892. But the last member of the family of noble gases, *radon*, a product of radioactive decay, had to wait another two years for isolation of a minute sample. When finally this whole corps was assembled, physical chemists and chemical physicists had six new reasons to agree that a new age of unity in the exact sciences had arrived. The advent of the atom as a respectable concept in physics as well as chemistry, however, was not so easily accomplished.[45]

Well into the twentieth century there were many efforts to codify separately a physics of matter and a physics of the aether. Mach, Ostwald, Helm, and Pierre Duhem (1861–1916) continued to propound their energetic anti-atomistic views until the evidence became simply overwhelming around 1910. Within chemistry a group of molecular theorists, supporting Ostwald's emphasis on energy transfer and Kelvin's speculations on vortex tetrahedral spheres, were urging the adoption of a colloidal conception of the aether.

Another effort to hybridize the physical sciences by casting a chemical foundation for a physical aether came from one of the chief architects of the Periodic Table. Dimitri Mendeleev (1834–1907), the grand old Russian whose intuition and taxonomy had borne such elaborate fruit in predicting the discovery and order of the elements, tried at the end of his life to fit the electromagnetic aether as another inert gas underneath hydrogen. His effort appeared in English in 1904, but it was stillborn, lost in the shuffle of still more basic philosophic issues.[46]

From a French perspective, the last decade of the nineteenth century appears to have been characterized more by subcultural arguments over the "bankruptcy of science" than by the so-called "energetics controversy" or debates about the reality of aether, electrons, and atoms. Henri Poincaré (1854–1912), then professor of mathematical physics at the Sorbonne, accomplished so much in celestial mechanics, electricity, optics, and pure mathematics during this period that he was clearly recognized as one of the foremost savants in all Europe. He could tackle any problem and find a new approach, it seemed; so his reputation as a universalist spread rapidly and far afield. But his character as a philosopher of science was not well known in the rest of Europe until his books began to appear in translations during the next decade. By then Poincaré seemed to assume the role of peacemaker between warring factions of physicists and philosophers exercised over the ontology of atoms and aether and the epistemology of kinetic-molecular and electromagnetic theories.[47]

Pierre Duhem became the theoretical physicist at Bordeaux in 1895, and from basic studies in fluid mechanics, chemistry of liquids, electro- and thermodynamics he progressed to analyses of disciplinary and national styles in science at large. Duhem was sympathetic toward Mach and Ostwald, antipathetic toward mechanism, materialism, and corpuscular theories. Like Poincaré, he tried to redefine the goals and methods of science in terms of distinctions between descriptions and explanations, models and analogies, science and metaphysics. Unlike the master mathematician, Duhem had a historical and sociological conscience.

Duhem's hostility toward "atomistics" was widely shared, but his sympathy toward energetics was not concomitantly shared. It seems as if Boltzmann was more the *bête noire* than Ostwald in the 1890s. Marcel Brillouin (1854–1948), another hydrodynamicist and fluid mechanician at the Ecole Normale, was one of the few French professors who thoroughly appreciated the Maxwell-Boltzmann doctrine of statistical mechanics. His protégé Jean Perrin (1870–1942), who published his first experimental results in 1895 before obtaining his doctorate in 1897 with more evidence on cathode rays and X rays, came to devote most of his life to proving "molecular reality." Together with Paul Langevin (1872–1946), freshly returned from studying with J. J. Thomson at Cambridge, Perrin moved into physical chemistry, then into the study of thermodynamics, colloids, Brownian movement, and the eventual corroboration of molecular, atomic, electronic, and nucleonic reality.[48]

Max Planck and Wilhelm Wien in Germany were slowly and painfully moving in the same direction. In 1898, Wien published his proofs that canal rays behind a perforated cathode are positively charged. This report complemented the work of J. J. Thomson quite nicely, except that positively charged particles were so much larger. Following the furrows of Lenard and Röntgen, Wien provided confirmations of Perrin's first reports, definitions, and data for Planck's deliberations. Wien's 1911 Nobel Prize address, "On the Laws of Thermal Radiation," perfectly adumbrated Planck's 1920 Nobel Prize lecture, "The Genesis and Present State of Development of the Quantum Theory." Wien's energy-distribution law provoked Planck to derive the relationship that energy emitted by a resonator must be discontinuous ($E = h\nu$).[49]

Guglielmo Marconi (1874–1937), a privileged son of an Italian gentleman and an Irish lady, began experiments with Hertzian waves for wireless signals on his father's estate near Bologna in 1895. Within the year, he had invented a practical system of wireless telegraphy that worked over a distance of a mile and a half. In September 1896, Marconi demonstrated his system of signaling for the British Government on Salisbury Plain first and then across the Bristol Channel. In 1897, he formed the Wireless Telegraph and Signal Company, Ltd., and by March 1899, Marconi's company had established wireless communication across the English Channel from near Dover, England, to near Boulogne, France. Most physicists felt that this distance (approximately 50 km) was about the limit to be expected from Hertzian waves, because they are propagated in straight lines and thus should move away from the curvature of Earth. But young Marconi, undaunted, changed the name of his company to his own, applied for a patent on a complete system of transmitters and receivers called "tuned or syntonic telegraphy," and was issued British patent number 7777 on 26 April 1900.

Carl F. Braun (1850–1918), professor and principal of the Strasbourg Physics Institute, who had invented a cathode-ray oscilloscope in 1897, turned his attention in 1898 toward Marconi's need for a stronger transmitter so as to reduce the need for larger antennae. Braun succeeded by 1900 in devising "sparkless telegraphy." Thanks to Braun's work at the Strasbourg forts and Marconi's efforts with condenser circuits, long-distance radio telegraphy was born. In January 1901, Marconi sent and received signals from two points along the south coast of England 186 miles apart. On 12 December that year, he succeeded in signaling across 2,100 miles of the Atlantic Ocean between Cornwall and Newfoundland. For these achievements Marconi and Braun shared the 1909 Nobel Prize.[50]

The approach of the year 1900 evoked floods of comment on the past progress and future prospects for science and technology in remaking man's world. Both professorial and popular writers of all sorts gave testimonials to past wonders wrought and future marvels sought. Of all the many such products and projects of minds and hands, none seemed more wonderful and marvelous than the electromagnetic aether of space. Sending messages, and maybe later voices and pictures, through empty space at the speed of light—that must surely mean that the vacuum is a plenum after all, and that energy transformations are more fundamental than material transformations. Jules Verne and H. G. Wells could hardly suggest anything that was totally preposterous any more, because it seemed that few things, if any, were impossible. Indeed, it was beginning to seem as if a primary function of the science of the future should be to show what was impossible and why, to discover more principles of impotency like the second law of thermodynamics.

Around Gustave Eiffel's 300-m tower, built as the tallest structure in the world for the International Exhibition of 1889, another International Exposition was held in 1900, which featured among many more spectacular novelties (including Count Zeppelin's first dirigible) an International Congress of the Sciences. Poincaré and Kelvin were featured speakers in the physics section of this affair, and both were more proud than humble over the achievements of physical science, past and future. Poincaré spoke on "Aether and Matter" in response to Larmor's new book with that title. He here first postulated that the speed of light should be recognized as a new absolute limit for natural velocities. Kelvin addressed himself to long-standing similar concerns, which he later characterized as two "Nineteenth Century Clouds over the Dynamical Theory of Heat and Light." The first "cloud" concerned the relative motion of aether and ponderable bodies and culminated with the Michelson-Morley experiment, which seemed to him impeccable and therefore led to his judgment that this cloud was still "very dense." The second "cloud" to which Kelvin gave even more attention was the Maxwell-Boltzmann doctrine of the partition of kinetic energy. Kelvin and Rayleigh both felt a fundamental difficulty with the received kinetic-molecular theory of gases. When applied to radiation, the "law" of equal partition seemed to disregard potential energy. This cloud had obscured the brilliance of the dynamic or molecular theory of heat and light since at least 1875. It now seemed to threaten a thunderstorm of reappraisals of such basic concepts as what are meant by "rigid bodies," "degrees of freedom," "point forces," and "center of inertia." Perhaps other concepts more basic still, such as space, time, aether, atom, molecule, ion, and electron, would likewise have to be revised.[51]

London's Royal Institution celebrated its centenary in June 1899, and on that occasion Rayleigh praised Thomas Young for having coined almost a hundred years earlier the concept "energy" that had become the dominant motif of physical sciences. Gerald Molloy, a Royal Institution lecturer, later encapsulated the essence of the lessons learned from aether-waves as follows:

Thus we come to realize that the various forms of energy to which in common language, we give the names of electricity,

magnetism, heat, light, chemical action, are all transmitted
through space in the form of waves or vibrations of the ether;
and that these vibrations are all essentially of the same kind,
being distinguished from one another only by their wave-
length. They produce widely different effects, when they
strike upon our different senses; but considered in themselves
they are only, so to say, notes of different pitch in the great
scale of radiant energy. . . . This large and comprehensive
view of radiant energy is one of the most notable results
achieved by the great scientific men of the century that has
just passed away. And the work that remains to be done by
the coming generation . . . is to explore more thoroughly the
properties of these ethereal waves, to fill up the gaps that still
exist in the scale, and perhaps to reveal to the world the intrinsic constitution of the ether itself, that mystery of mysteries which underlies all those outward phenomena of nature.[52]

This attitude toward the aether as the "mystery of mysteries"
was echoed over and over again at the turn of the century. Eminent physicists of all sorts were wrestling with ontological
questions of the reality of aether, electrons, and atoms. Although
Kelvin was assailed by self-doubt stemming largely from the
Michelson-Morley failure to detect an aether wind, R. A. Fessenden, among others encouraged by Hertz and his successors, published a new determination of electromagnetic quantities, which
supposedly quantified the "density and elasticity of the ether."
Michelson himself and H. A. Rowland, Thomas Preston, Paul
Drude, Osborne Reynolds, and G. F. FitzGerald were among those
who argued in various ways for the electromagnetic aether to be
recognized as the primary property of the physics of the twentieth
century.[53]

If Larmor, Poincaré, and Lorentz seemed to be on the verge of
solving the main problems of the interactions between aether, matter, and energy transfer, the electrification of the scientific world-
view might supersede the mechanization and materialism associated with Newtonian assumptions. Even the most famous of
the newest advocates of discontinuity in nature, J. J. Thomson and

Max Planck, were extremely cautious about allowing themselves to think against the tide of continuity.

By 1900, the last year of a fantastically progressive century, especially in science and technology, the artificial world of man's creation was beginning to rival the variety and uniqueness of Earth itself. To 60 million or so visitors to the great International Exhibition in Paris that year, urban, industrial cultures must indeed have seemed bound for the conquest of all nature. The passing century had witnessed the girdling of Earth by telegraph and submarine cables, steam locomotives and steel ships, such engineering works as the Suez Canal, the Firth of Forth Bridge, and transcontinental railroads. Electric dynamos, motors, light and power systems, and countless electric gadgets were proliferating throughout the world, as were telephones, bicycles, automobiles, sewing machines, typewriters, rotary printing presses, photography, phonography, and mechanical implements for agriculture. Chemistry had introduced a new age of synthetic materials of all sorts, and biology had contributed heavily toward improvements in food, drink, medicine, surgery, sanitation, and public health. Physics had learned from steam engines how to reverse processes of heat transfer and build ice machines capable of liquefying air and freezing gases. Röntgen's X rays were the latest rage, but physical optics had so improved microscopy, telescopy, spectroscopy, and interferometry that one U.S. Patent Office examiner could say in 1900 that "to the sum total of human knowledge no department has contributed more than that of optics." An age of cheap steel and rock oil had arrived. An apparently endless Industrial Revolution must almost inevitably breed intellectual revolutions in its wake.

NOTES

1. For a standard guide to the background of our titular concerns here, see Sir Edmund Whittaker, *A History of the Theories of Aether and Electricity*, 2 vols. (New York: Harper Torchbooks, 1960); originally London, 1910, rev. and enlarged 1951 (vol. I), 1953 (vol. II). For a helpful anthology, see Henry A. Boorse and Lloyd Motz, eds., *The World of the Atom*, 2 vols. (New York: Basic Books, 1966).
2. Max Talmy, M.D., *The Relativity Theory Simplified and the Formative*

Period of Its Inventor (New York: Falcon Press, 1932), p. 142. Aaron Bernstein, *Naturwissenschaftliche Volksbücher,* 21 vols. (Berlin: G. Hempel, 1880). For two of the more reliable biographies for this period of his life, see Rudolf Kayser, a son-in-law writing under the pseudonym "Anton Reiser," *Albert Einstein: A Biographical Portrait* (New York: A. & C. Boni, 1930), and Carl Seelig, *Albert Einstein: A Documentary Biography,* trans. Mervyn Savill (London: Staples Press, 1956).

3. Ludwig Büchner, *Force and Matter: or Principles of the Natural Order of the Universe with a System of Morality Based Thereon,* trans. from 15th German edn. by author (New York: Truthseeker, 1913).

4. See Gerald Holton's "Influences on Einstein's Early Work" in his *Thematic Origins of Scientific Thought: Kepler to Einstein* (Cambridge: Harvard Univ. Press, 1973), pp. 197–217; Ronald W. Clark, *Einstein: The Life and Times* (New York: World, 1971), pp. 15–23. See also Jagdish Mehra, "Albert Einstein's 'First' Paper," *Science Today,* April 1971.

5. Albert Einstein, "Autobiographical Notes," in Paul A. Schilpp, ed., *Albert Einstein: Philosopher-Scientist,* 2 vols. (New York: Harper Torchbooks, 1959), vol. I, pp. 15, 19, 21, 33, 53.

6. For Lodge, see the entry by Charles Süsskind in C. C. Gillispie, ed., *DSB* (see note 3 of Chap. 1, above), vol. VIII, pp. 443–44. For Lenard, see the entry by Armin Hermann in Gillispie, *DSB,* Vol. VIII, pp. 180–83.

7. On the roles of Philipp Lenard and Johannes Stark in Nazi physics, see Joseph Haberer, *Politics and the Community of Science* (New York: Van Nostrand Reinhold, 1969), pp. 103–84. For Lenard's orientation, see his *Great Men of Science: A History of Scientific Progress,* H. S. Hatfield, trans. (New York: Macmillan, 1933).

8. Sir Oliver Lodge, *Modern Views of Electricity,* 2nd edn. (London: Macmillan, 1892), Preface to 1st edn. (1889), p. vii. The answering quotation comes from an appendix in this text from an 1882 lecture by Lodge (see p. 416). For Lodge's orientation, see David B. Wilson, "The Thought of Late Victorian Physicists: Oliver Lodge's Ethereal Body," *Victorian Studies* 15 (September 1971): 29–48.

9. Lodge's report on present state of knowledge with respect to aether and matter in *Minutes* of Physical Society, meeting 27 May 1892, in *Nature* 46 (16 June 1892): 164–65.

10. Geo. Fras. FitzGerald, "The Ether and the Earth's Atmosphere," *Science* 13 (17 May 1889): 390.

11. George Johnstone Stoney, "On the Cause of Double Lines and of Equi-Distant Satellites in the Spectra of Gases," *Royal Dublin Society, Trans.* 4 (1891): 583. Italics in original.

12. See the discussion and selections in Kenneth F. Schaffner, *Nineteenth-Century Aether Theories* (Oxford: Pergamon, 1972), pp. 76–98, 194–207, *Re* the Anglo-Irish tradition, see the two essays by Sir Edmund Whittaker on Sir William Rowan Hamilton and G. F. FitzGerald in the *Sci-*

entific American book *Lives in Science* (New York: Simon and Schuster, 1957), pp. 59–83.

13. Oliver Heaviside, *Electromagnetic Theory*, 2 vols. (London: "The Electrician" Publishing Co., 1893, 1899), vol. I, p. x; Rollo Appleyard, *Pioneers of Electrical Communication*, op. cit., has a good chapter on Heaviside, pp. 211–60. *Re* Kelvin and Stokes, see their addresses of 1892 and 1893, respectively, in Sir Wm. Thomson [Kelvin] *Popular Lectures and Addresses*, 3 vols. (London: Macmillan 1894), vol. II, pp. 535–56; and Stokes, "The Luminiferous Aether," Smithsonian Institution, *Annual Report, 1893* (Washington, D.C., 1893), pp. 113–19. Also, Lord Salisbury, *Evolution: A Retrospect* (London: Roxburghe Press, 1894), pp. 28–29.

14. J. J. Thomson, *Notes on Recent Researches in Electricity and Magnetism* (Oxford: Clarendon Press, 1893), p. 2. *Re* J. J. Thomson's attitudes toward the aether and toward his elders' attitudes, see Sir J. J. Thomson, *Recollections and Reflections* (New York: Macmillan, 1937), pp. 48–100, 237–41, 369, 394, 432; and George P. Thomson, *J. J. Thomson: And the Cavendish Laboratory in His Day* (London: Nelson, 1964), pp. 4–13, 37, 155–57.

15. R. T. Glazebrook, *James Clerk Maxwell and Modern Physics* (New York: Macmillan, 1895), pp. 204–5; cf. Glazebrook's presidential address to the British Association for the Advancement of Science (BAAS) (Section A), "Review of Optical Theories," BAAS, *Annual Report, 1893*, pp. 673–81. *Re* these and other such difficulties, especially with notation problems (vectors vs. quaternions), see Alfred M. Bork, "Physics Just Before Einstein," *Science* 152 (29 April 1966): 597–603.

16. Tetu Hirosige, "Origins of Lorentz' Theory of Electrons and the Concept of the Electromagnetic Field," in Russell McCormmach, ed., *Historical Studies in the Physical Sciences*, vol. I (1969), pp. 151–209. See also McCormmach's article in vol. II (1970), pp. 41–87, "Einstein, Lorentz, and the Electron Theory," and his introductions to each of these volumes, as well as his article "H. A. Lorentz and the Electromagnetic View of Nature," *ISIS* 61 (Winter 1970): 459–97.

17. On Lenard, see Nobel Foundation, *Nobel Lectures*, including Presentation Speeches and Laureates' Biographies, *Physics, 1901–1921* (Amsterdam: Elsevier, 1967), pp. 100–138 (hereafter cited as *Nobel Lectures Physics*).

18. See especially Robert S. Cohen's introductory essay to the Dover edition of Heinrich Hertz, *The Principles of Mechanics Presented in a New Form* (New York: Dover, 1956), and Hertz's Introduction, p. 18.

19. Ibid., p. 144.

20. Ibid., pp. 27, 38, 41.

21. Helmholtz's Preface in ibid., pp. xxxii–xxxiii.

22. See Lenard's lecture in *Nobel Lectures Physics* (1905), p. 108.

23. Paul Drude, *Physik des Aethers auf Elektromagnetischer Grundlage*

(Stuttgart: F. Enke, 1894); August Föppl, *Einführung in die Maxwellische Theorie der Elektrizität* (Leipzig: B. G. Teubner, 1894). See also Joan Bromberg's article on Föppl in Gillispie, *DSB*, vol. V, pp. 63–64.

24. Gerald Holton has made this translation as well as a strong case for Föppl's influence in his "Influences on Einstein's Early Work," *Thematic Origins of Scientific Thought*, pp. 205–9.

25. For Wien, see *Nobel Lectures Physics* (1911), pp. 269–89.

26. For Boltzmann, see Stephen G. Brush's entry in Gillispie, *DSB*, vol. II, pp. 260–68; see also Brush's *The Kind of Motion We Call Heat: A History of the Kinetic Theory of Gases in the 19th Century* (Studies in Statistical Mechanics, vol. VI), 2 books (Amsterdam: North-Holland, 1976). For some of the main difficulties encountered with this primitive concept even now, see D. W. Theobald, *The Concept of Energy* (London: E. & E. N. Spon, 1966).

27. Among the library of works on Mach, see especially John T. Blackmore, *Ernst Mach: His Life, Work and Influence* (Berkeley: Univ. of California Press, 1972), and J. Bradley, *Mach's Philosophy of Science* (London: Athlone Press, 1971).

28. Max Planck, *Scientific Autobiography*, Frank Gaynor, trans. (New York: Philosophical Library, 1949), esp. pp. 13–22, 33–34.

29. See Niles R. Holt, "A Note on Wilhelm Ostwald's Energism," *ISIS* 61 (Fall 1970): 386–89.

30. On the Lübeck controversy and its significance, see Erwin N. Hiebert, "The Energetics Controversy and the New Thermodynamics," with commentaries by Lawrence Badash and David B. Wilson, in Duane H. D. Roller, ed., *Perspectives in the History of Science and Technology* (Norman: Univ. of Oklahoma Press, 1971), pp. 67–97.

31. For Lorentz, see Russell McCormmach's entry in Gillispie, *DSB*, Vol. VIII, pp. 487–500. For Larmor, See A. E. Woodruff's entry in Gillispie, *DSB*, Vol. VIII, pp. 39–41.

32. Sir Joseph Larmor, *Aether and Matter: A Development of the Dynamical Relations of the Aether to Material Systems on the Basis of the Atomic Constitution of Matter Including a Discussion of the Influence of the Earth's Motion on Optical Phenomena* . . . (Cambridge: Cambridge Univ. Press, 1900). See also Barbara G. Doran, "Joseph Larmor and the Roots of Modern Physics," unpublished Ph.D. thesis, Johns Hopkins University, 1973.

33. H. A. Lorentz, *Versuch einer Theorie der electrischen und optischen Erscheinungen in bewegten Körpern* (Leiden: E. J. Brill, 1895), reprinted Leipzig, 1906, and in H. A. Lorentz, *Collected Papers*, 9 vols., P. Zeeman and A. D. Fokker, eds. (The Hague: M. Nijhoff, 1935–39), vol. V, pp. 1–137.

34. Lorentz in *Nobel Lectures Physics* (1902), p. 19, and, for next quotation, p. 20.

35. Larmor, *Aether and Matter*, p. vi, note, and p. 78.

36. Röntgen was the first Nobel laureate in physics, but he delivered no acceptance lecture: see *Nobel Lectures Physics* (1901), pp. 3–8 for presen-

tation speech and biography. See also W. Robert Nitske, *The Life of Wilhelm Conrad Röntgen; Discoverer of the X-Ray* (Tucson: Univ. of Arizona Press, 1971).

37. This translation by George F. Barker is reprinted in W. F. Magie, ed., *A Source Book in Physics* (Cambridge: Harvard Univ. Press, 1935), pp. 601, 607; cf. Röntgen, "On a New Form of Radiation," *Nature* 53 (1896): 274.

38. Becquerel and the Curies shared the Nobel prize for 1903; for their interactions, see *Nobel Lectures Physics* (1903), pp. 45–83.

39. For two excellent guides to the history of subsequent events, see Alfred Romer, *The Discovery of Radioactivity and Transmutation* (New York: Dover, 1964) and *Radiochemistry and the Discovery of Isotopes* (New York: Dover, 1970).

40. P. and M. S. Curie, in Magie, *A Source Book in Physics*, p. 615.

41. J. J. Thomson, "Cathode Rays," *Philosophical Magazine*, 5th series, 44 (October 1897): 293–316; reprinted in Magie, *Source Book in Physics*, op. cit., p. 589.

42. J. J. Thomson, "On the Masses of the Ions in Gases at Low Pressures," *Philosophical Magazine*, 5th series, 48 (December 1899): 548. Among J. J. Thomson's research students at the Cavendish in 1898 were C. T. R. Wilson, Paul Langevin, O. W. Richardson, Ernest Rutherford, W. C. Henderson, J. Zeleny, G. B. Bryan, and H. A. Wilson.

43. See A. S. Eve, *Rutherford* (London: Cambridge Univ. Press, 1939). See also David L. Anderson, *The Discovery of the Electron: The Development of the Atomic Concept of Electricity* (Princeton: Van Nostrand, 1964).

44. On Rayleigh, see *Nobel Lectures Physics* (1904), pp. 85–98. On Ramsay, see *Nobel Lectures Chemistry* (1904), pp. 65–79.

45. William Ramsay, *The Gases of the Atmosphere: The History of their Discovery*, 3rd edn. (New York: Macmillan, 1905).

46. D[mitri] Mendeléef, *An Attempt towards a Chemical Conception of the Ether*, George Kamensky, trans. (London: Longmans, Green, 1904); cf., Carl Barus, "The Compressibility of Colloids, with Applications to the Jelly Theory of the Ether," *American Journal of Science*, 4th series, 6 (October 1898): 285–98.

47. On Poincaré, see Tobias Dantzig, *Henri Poincaré: Critic of Crisis* (New York: Scribner's, 1954). For Poincaré's lectures at the Sorbonne on Maxwell-Helmholtz problems during the 1888, 1890, and 1899 terms, see his *Electricité et optique*, 2nd ed. (Paris: Gauthier-Villars, 1901).

48. Mary Jo Nye, *Molecular Reality: A Perspective on the Scientific Work of Jean Perrin* (New York: American Elsevier, 1972). For Pierre Duhem, see the entry by Donald S. Miller in Gillispie, *DSB*, Vol. IV, pp. 225–33.

49. For Planck, see *Nobel Lectures Physics* (1918), pp. 403–20. For Wien, see ibid. (1911), pp. 269–90.

50. On Marconi and Brown, see *Nobel Lectures Physics* (1909), pp. 191–248. See also George Shiers, "The First Electron Tube," *Scientific American*

220 (March 1969): 104–12, with comments by Lloyd Espenschied, *Scientific American* 221 (August 1969): 8; and Shiers, "Ferdinand Braun and the Cathode Ray Tube," *Scientific American* 230 (March 1974): 92–101.

51. See Richard D. Mandell, *Paris 1900: The Great World's Fair* (Toronto: Univ. of Toronto Press, 1967); Edward W. Byrn, *The Progress of Invention in the Nineteenth Century* (New York: Munn, 1900); Henri Poincaré, *Aether and Matter* (Paris: Gauthier-Villars, 1900); Kelvin, "Nineteenth Century Clouds over the Dynamical Theory of Heat and Light," reprinted as Appendix B-10 in rev. edn., *Baltimore Lectures on Molecular Dynamics and the Wave Theory of Light* (London: C. J. Clay & Sons, 1904).

52. Gerald Molloy, "Electric Waves," a Friday Evening Discourse at the Royal Institution on 15 February 1901, in William L. Bragg and George Porter, eds., *The Royal Institution Library of Sciences . . . Physical Sciences,* vol. 5 (New York: American Elsevier, 1970); p. 371; cf. pp. 277, 324, 551.

53. R. A. Fessenden, "A Determination of the Nature of the Electric and Magnetic Quantities and of the Density and Elasticity of the Ether," *Physical Review* 10 (January 1900): 1–33, and 10 (February 1900): 83–115. A. A. Michelson, *Light Waves and Their Uses* [Lowell Lectures, 1899] (Chicago: Univ. of Chicago Press, 1903); Henry A. Rowland, "The Highest Aim of the Physicist," *American Journal of Science*, 4th series, 8 (December 1899): 401–11; Thomas Preston, *The Theory of Light*, 3rd edn., C. J. Joly, ed. (London: Macmillan, 1901); Paul Drude, *The Theory of Optics*, C. R. Mann and R. A. Millikan, trans. (New York: Longmans, Green, 1901); Osborne Reynolds, *The Sub-Mechanics of the Universe*, vol. III of *Papers on Mechanical and Physical Subjects* (Cambridge: Univ. Press, 1900–1903). George Francis FitzGerald, *The Scientific Writings of the Late George Francis FitzGerald*, Joseph Larmor, ed. (London: Longmans, Green, 1902), pp. 511, 514.

Chapter IV

Analyses of Relativity
(c. 1900s)

Whether the ether exists or not matters little—let us leave that to the metaphysicians: what is essential for us is that everything happens as if it existed. . . .

Someday, no doubt, the ether will be thrown aside as useless.

—Henri Poincaré, 1902

We will raise this conjecture (the purport of which will hereafter be called the "Principle of Relativity") to the status of a postulate, and also introduce another postulate . . . namely, that light is always propagated in empty space with a definite velocity c which is independent of the state of motion of the emitting body. These two postulates suffice for the attainment of a simple and consistent theory of the electrodynamics of moving bodies based on Maxwell's theory for stationary bodies. The introduction of a "luminiferous ether" will prove to be superfluous inasmuch as the view here to be developed will not require an "absolutely stationary space" provided with special properties, nor assign a velocity-vector to a point of empty space in which electromagnetic processes take place.

—Albert Einstein, 1905

Einstein simply postulates what we have deduced. . . . By doing so, he may certainly take credit for making us see in the negative results of experiments like those of Michelson, Rayleigh and Brace, not a fortuitous compensation of opposing effects, but the manifestation of a general and fundamental principle.

Yet, I think, something may also be claimed in favour of the form in which I have presented the theory. I cannot but regard the ether . . . as endowed with a certain degree of substantiality, however different it may be from ordinary matter.

—H. A. Lorentz, 1909

145

Twentieth-century relativity theory was a collective accomplishment, not merely a singular confrontation of a solitary genius with the universe. A small, informal group of European professors of physics, fewer than a thousand men altogether but certainly more than a dozen, wrestled over several decades at the turn of the century with problems concerning certain confusions between electricity and electronics, problems relating to both atomic physics and astrophysics. By trying to understand phenomena so far beneath the scale of ordinary human experience and above the level of ordinary imagination, these physical theorists departed from levels of common discourse. Searching for better deductive languages from mathematics, wherein the study of relationships could reach beyond the powers of natural words and syntax, many physical scientists became mathematical physicists, a subtle change with profound implications, because they came to behave as a breed apart from their philosophical community.

Those who stressed mathematical modeling over physical experiencing as the most adequate way to describe or explain and predict the behavior of physical nature were often accused of being "phenomenalists" because they seemed to avoid philosophical commitments to a theory of reality. On the other hand, those who stressed experience over theoretical models became labeled by their rivals as "neo-positivists," because, like the Compteans long before, they sought to believe only in positively verifiable objectivity. Despite their scholarly rites and academic rituals, however, most scientists, whether "positivistically" or "phenomenalistically" inclined, were no more exempt than other men from certain kinds of pride and prejudice, sensitivity and sensibility; they had certain roles to play and protocols to observe. They struggled to understand events in contexts so exponentially large and small that often even the disciplined hands and minds of their colleagues in neighboring disciplines could not fully grasp the meanings of their superspecialized jargon. Growing professional awareness of the infinite complexity of the world seemed surpassed only by the eternal mysteries surrounding man's consciousness of it.[1]

Four men in particular stood in the vanguard of physics and philosophy at the turn of the century. Each of them was a leader in the analyses of those problems that led into the theory of relativity.

Each was also in one sense a progenitor of some aspect of that theory or its larger context. Ernst Mach (1838–1916) of Vienna was still, as both physicist and philosopher, extremely influential in the movement to purge all metaphysical notions from scientific thought and practice. As a result he became widely known as the father of epistemological relativity. Hendrik A. Lorentz (1853–1928) of Leyden was the foremost theoretical physicist, whose electron theory by 1900 was making the field concept appear more real than matter. He was already widely revered as the founder of the electromagnetic world-view. J. Henri Poincaré (1854–1912) of Paris had achieved status as a savant among mathematicians, astronomers, and physicists during the 1880s and 1890s for the great breadth and profundity of his contributions to knowledge. Poincaré had coined the phrase "principle of relativity" to stand for science's failure to determine Earth's absolute motion, and he was therefore widely regarded as a rival of Mach's for the title of father of relativity theory. And finally Max Planck (1858–1947) of Berlin was the theoretical physicist whose analyses of radiation had led him reluctantly, late in 1900, to a radical solution for the so-called black-body problem of energy transfer. For this and subsequent reinforcements of analyses in subatomic physics, Planck soon became widely revered as the father of the quantum of action. By 1901, Mach of Austria, Lorentz of Holland, Poincaré of France, and Planck of Germany were recognized molders and shapers of opinion in European physics.

Whereas many others before 1905 might also be listed as progenitors of modern relativity theory, Albert Einstein should not be among them. For this young scientist evolved into the position later assigned him as the father of modern relativity. In 1905, he did not overtly propose a "revolution" or confront orthodoxy with a demand for an overhaul of physics. Rather, he modestly proposed a basic pair of postulates for a different view of classical electrodynamics as he understood it. It is true that his different view did include a modification of the theory of space and time, which were basic physical concepts. Yet his primary efforts at first were to preserve the invariances and covariances between the coordinate systems of material points or rays of light so that theoretical order could be imposed on systems in relative motion. In communicating what he and no one else quite so fully understood, however, Einstein came to deserve credit for paternity, especially because his other papers on physical problems were equally bold, complementary, and almost as influential.

Einstein's Growth—To 1905

In August 1900, Albert Einstein graduated with a diploma in mathematics and science from the Swiss polytechnic institute, the ETH in Zurich. He was twenty-one years old and eager to find a place in an academic setting where he could continue his studies into the fundamental problems of physics and chemistry. But having renounced his native citizenship and not having yet acquired Swiss papers, although he had applied a year earlier, young Einstein found himself with very poor prospects for an academic career. While visiting his parents in Milan, he began seeking employment by mail, because he had failed to gain a fellowship at his alma mater. Over the next year he did some tutoring and substitute teaching, but his applications for graduate work in the laboratories of Wilhelm Ostwald, Heike Kamerlingh Onnes, and other famous scientists were all ignored.[2]

In February 1901, Einstein was granted Swiss citizenship, and although job prospects remained bleak, he polished up a paper on capillarity phenomena, submitted it to *Annalen der Physik*, and was pleasantly surprised to find it accepted by the editor, Paul Drude. When the paper was published in the December 1901 issue of this prestigious journal, Einstein secured reprints to show and send with his job applications. Old friends from the ETH, especially Marcel Grossmann and Conrad Habicht, helped him celebrate these high points of an otherwise dry year in Zurich. His father had died that year, and his mother and his sister Maja were in need of support, but still no job appeared.

Meanwhile, Einstein took advantage of Drude's receptivity to send off two more brief articles, one on thermodynamics and another on kinetic theory, both of which were published in 1902. These publication victories helped restore Einstein's confidence, for his pride had been sorely tested during his employment search. He felt a certain paranoia toward his old physics professor Heinrich Weber, but the ETH historian Alfred Stern had befriended him. And now Marcel Grossmann's father was helping to influence the director of the Swiss Patent Office, Herr Friedrich Haller, to interview Einstein for a post in Berne. After applying for the title and salary of a "technical expert (second class)," he was hired in June 1902 but as a "technical expert (third class)." The job was for an electromechnical engineer to examine and criticize patent applications and specifications.

Einstein's move from Zurich to Berne was a fateful change in many ways. As a young leftist Jewish student in the counterculture of Zurich, Einstein participated in the cosmopolitan and intellectual life of a city that was at the crossroads of European revolutionary ferment. Socialists, nihilists, Zionists, and sectarians of all sorts—including Lenin, Plekhanov, Mussolini, Rosa Luxemburg, Theodor Herzl, Carl Jung, Friedrich Adler—congregated in Zurich and other Swiss cities nearby to lay their plans for, they hoped, a better twentieth-century world. If "generational conflict" explains much of the sociological background to Einstein's maturation as a philosophical rebel, his psychological development as a creative revolutionary needed a focus and a discipline that were better served in Berne than in Zurich. By moving to the capital city of Switzerland and being employed as a civil servant, Einstein compromised himself enough to settle down to the hard task of reconciling his ideals with socio-political realities.[3]

Einstein had undoubtedly been somewhat sullen and lazy as a student, but in the context of European educational attitudes at the time, he always considered it surprising that he survived the system at all. He not only educated himself despite the authoritarian system, but he also imbibed a taste for intellectual adventure that led him ever deeper into the basic physical and philosophical problems of his day. His sensitivity to those problems and to subjective-objective distinctions in nature soon led him to ponder long and hard why certain mathematical formalisms seemed so artificial compared with how nature behaved. Fluctuations in radiation pressure were to him one prime example of such an intriguing problem.

Before landing the mechanical engineer's job at the Patent Office,

Einstein well knew he wanted primarily to be a physicist. But at that time he still might have gone toward either experimental or theoretical work, if other circumstances had prevailed. It happened that the patent examiner's task he acquired dealt mostly with specifications for electromotive devices and was not demanding or difficult. He had time for theorizing and reading much philosophy and physics.

By 1902, the house of physics was thoroughly divided between the theoretical and experimental sides, and there were pecking orders among theorists as well as among the ingenious inventors of new scientific and technological instruments. Elitist colleges, prestige publications, professional politics, and intellectual fashions all played some part in Patent Office affairs, as well as in the analyses of fields and particles that were taking place at the beginning of the century. Any neophyte entering upon a career in the physical sciences would do well to have some guidance around these professional obstacles, and yet too much concern with such matters might immobilize a creative talent.[4]

Einstein had to wrestle with such considerations (as well as with some anti-Semitism) for several years before and after 1900, but he preferred to immerse himself in thought and reading about natural philosophy. Although it was professionally inadvisable for budding young scientists to "waste" much time with philosophy, some students survived the specialization process despite the system. Young Einstein was one of those who aspired to be more than a mere mathematician or physicist or scientist or even merely an academic philosopher: he chose to try to understand nature. No single specialized discipline seemed sufficient to this quest anymore. Consequently, Einstein established his own priority reading lists and immersed himself in scientific classics. Hume, Kant, Mach, and Maxwell, followed by Hertz, Poincaré, Lorentz, Planck, and Ostwald's *Klassiker* collection, proved satisfying to his thirst for philosophy as well as physics. Most of his undergraduate teachers, however, had been German-speaking professors, such as the mathematicians Adolf Hurwitz and Hermann Minkowski, who were as puzzled as he was about the problems posed by Mach, Lorentz, Poincaré, and Planck. The graduate student knew that these famous names and others such as Boltzmann, Kaufmann,

Föppl, Lenard, and Wien constituted the guild of analyzers of physical phenomena in the scientific vanguard. Toward such men with high reputations Einstein had to look for judgment and approval if he hoped to make some contribution to solving the foremost problems with which they were concerned.

Curiously, most of those professing to know all or something about mathematical physics after 1900 were first-generation academic teachers who ignored or had forgotten the history of their own origins. Many had been trained first as engineers and then sought status through science. Striving toward leadership in their disciplines, these men felt no need to look back in time at the recency of their professional acculturation. Being future-oriented and generally contemptuous of the sociology and history of their discipline's accumulation of knowledge, they tended to ignore the temporal dimensions of their background and problems in favor of structural analyses, which seemed so much more tractable to their methods. Their textbooks stressed principles and problem-solving rather than conceptual understanding or challenging frontiers.

Once settled in his new job as a critical reader of technical specifications and drawings, Einstein was happy to be salaried and to have time to think about physics. His interests in statistical and analytical mechanics were being applied toward the theory of gas dynamics and molecular attraction, which seemed already in mid-1901 to be an appropriate subject for a doctoral thesis. Einstein continued to listen and look for possible academic posts to follow his employment as a civil servant. He never intended to remain indefinitely rewriting other people's patent papers.

Einstein also continued on the side as a tutor, offering private lessons in physics to whomsoever should respond to his want ads in the newspapers. Berne was a university city, and Maurice Solovine ("Solo") was a young Rumanian student who answered an ad just after Einstein arrived there. Soon the two were close friends, and they were joined by Conrad and Paul Habicht and by Michele-Angelo Besso ("Ange"), an Italian engineer, in an informal group of footloose graduate students who much enjoyed each other's company and conversations. They came to call themselves in jest the "Olympia Academy."[5]

On 6 January 1903, Albert Einstein married a former classmate and physics student, Mileva Marič. She was of Serbian extraction and had dared to invade the traditionally all-male discipline of physics. She was four years older than Albert, came from Hungary, and suffered from an imperfect leg. Mileva was apparently determined at first to be a personage in her own right. She loved Switzerland and enjoyed the male camaraderie of the "Olympians" during their first year together. Albert was exempt from military service because of flat feet and varicose veins. The couple began an idyllic marriage in 1903, with congenial friends, many picnics and excursions, and much simple conviviality. Then Mileva became pregnant and on 14 May 1904 gave birth to a boy, whom they named Hans Albert. Many years and many troubles later, he would become a professor of hydraulic engineering at the University of California (Berkeley). Eventually Mileva lost her spark to the three Wilhelmine K's of Germanic domestic life: *Kinder, Küche, und Kirche* (children, cooking, and church). But early in their marriage she was undoubtedly of great help to her husband.

Mileva, Solo, Ange, the Habicht brothers, and others occasionally acted as sounding boards for Albert, allowing him to test in conversation and argument ideas that could later be honed on paper. When Michele Besso also came to work at the Swiss Patent Office in 1904, Einstein began to enjoy a permanent and daily contact with a knowledgeable friend who could keep him honest in the exploration of physical ideas. Einstein and Besso together explored in talks and reading the current fads of European inventors and fashions of mathematical physics. But Einstein had the urge to get beyond superficialities by publishing his technical notions. Besso the engineer apparently had stronger social and economic interests. By now Einstein the physicist clearly wanted to obtain his doctorate and to find an academic post. In 1905, thanks to the sympathetic encouragement of Professor Alfred Kleiner, Einstein finally achieved his Ph.D. from the University of Zurich for a dissertation on "A New [Method for] Determination of Molecular Dimensions." This work was significant in its own right, but it was quickly overshadowed by Einstein's other works of that same year.

If the plague year of 1665–66, was the *annus mirabilis* for Isaac Newton's main contributions to science, the miraculous year for Albert Einstein was 1905. There appeared under his name in *Anna-*

len der Physik five papers that were written that year, and three of them were major contributions to theoretical physics in quite different ways. The first of these major papers, on the radiation of light and its energy transfer considered from Planck's quantum standpoint, Einstein himself evidently considered the most "revolutionary" at the time. The photoelectric effect discussed therein likewise impressed his colleagues long afterward as the most basic and least controversial, for it was to be ostensibly for this work especially that Einstein was to receive the Nobel Prize of 1921.[6]

The second paper, on the real dimensions of molecules that should be determinable from the random motions of suspended particles in solution, was closely connected with Einstein's doctoral dissertation. Brownian motion had long been a phenomenon known to biologists, but Einstein started a new trend in molecular physics with his suggestions for applying the kinetic theory of heat to the problems of statistical thermodynamics. Chemical physics and physical chemistry would find their marriage renewed eventually by the implications of these interrelationships. The physical reality of molecules and atoms was here reinforced.[7]

The third and most epochal of Einstein's three major papers of 1905 was to become the birth certificate of the theory of relativity. Entitled similarly to the problematic theses of Maxwell, Helmholtz, Hertz, J. J. Thomson, and H. A. Lorentz, among others, Einstein's article "On the Electrodynamics of Moving Bodies" was the work that would most enrapture young radicals in physics over the next several decades. Because its philosophical critique of Newtonian mechanics was in accord with its physical critique of the current situation in experimental and theoretical physics, Einstein's third paper of 1905 raised some fundamental issues that would eventually resound outside the profession. Modification of the theory of space and time seemed a most radical proposal, but at its roots this reformation was primarily an "effort to return to a classical purity."[8]

Einstein's paper, divided into two major sections (a kinematical part, and an electrodynamical part), was neither conceived nor first put forward as a "theory of relativity." It was, rather, first called informally by its author a "theory of invariance." And its purpose was merely to provide a professional solution to the technical problems of dealing with electrons and other very small energetic particles as if they were at rest rather than in perpetual dynamic flux. Einstein juxtaposed two most unlikely postulates in order to accomplish this simplification, but gradually the whole set of notions herein caught the attention, if not the admiration, of physicists, mathematicians, scientists generally, and then the general philosophical public. As we shall see later, the maturation of the 1905 paper from invariance theory into relativity theory, and thence into space-time physics, was an uncertain process that required many analyses over many years before Einstein's later syntheses began to seem, in many senses of the word, singular.

Machian Relativity

By common consent of both Einstein and the bulk of his supporters and disciples, the one most dominant intellect in the sciences at the turn of the century was Ernst Mach. Physicist, psychologist, philosopher, as well as physiologist and historian, Mach was the principal figure in central Europe around whom young rebels of all sorts could figuratively gather to express their independence of established dogmas. Mach's writings provided a corpus of criticism so broad and deep that no one wishing to reorganize the world could afford to ignore him.

Although his versatility and eminence made him a model Renaissance man, Mach was essentially a physicalist-missionary preaching for a new puritan reformation in science. Humanistic and moralistic, he like Karl Marx felt that the main tasks of science and philosophy were less to understand the world than to use and change it. Young activists of all persuasions found this stance attractive, but few could fully appreciate the subtleties that both Marx and Mach presented, whether in contrast or comparison. Marx, dead since 1883, had tried to establish a science of society on the basis of dialectical materialism. Mach, since the first edition of his *Science of Mechanics* (1883), had become the foremost epistemological relativist in Europe.[9]

Born on a farm in Moravia a day's journey from Vienna, Ernst Mach suffered through a lonely childhood full of frustrations before finding himself at the University of Vienna (1855–60) as a budding young physicist and mathematician. After earning his doctorate there and working as an unsalaried lecturer and experimentalist on a variety of physical and physiological problems, Mach went to Graz in 1864 for three years. There he found a wife to share his maturing interests in "psychophysics" and philosophy. In 1867, he was called to the chair of experimental physics at the University of Prague, where he remained until 1895. During these twenty-eight years, Mach served twice as rector of the university, became the father of four sons and a daughter, and produced a growing body of respectable original work, including the first photographs of the shock waves of projectiles (1886).

A family tragedy in 1894, the apparent suicide of a brilliant son just after his graduation with a Ph.D. degree from Göttingen, led to such grief and reticence in personal matters that the life of Ernst Mach was obscure until recently. But in 1895, Mach returned in triumph, despite this personal tragedy, to take the chair of philosophy at Vienna. Then, in 1898, he suffered a stroke that paralyzed the right half of his body, leading to his retirement in 1901. Nevertheless, he was that year also elevated to the Austrian House of Peers.

Incredibly persistent, energetic, and productive despite the afflictions of his old age, Mach became even more influential as new editions and translations of his works diffused through world markets. A leader of free-thinking intellectuals everywhere, Mach's iconoclastic ideas were valued more than his constructive views on the sciences. He defined science almost as a biological endeavor to economize on thought, and he used critical philosophy to try to strip away as much theory and as many primitive concepts as possible in favor of metrical and mathematical operations. Inveighing against absolutism in all forms, Mach assumed a role as a kind of high priest of an antimetaphysical inquisition. To him, aether, electrons, and atoms were all anathema.

Although Ludwig Boltzmann succeeded Mach in the chairs of philosophy and physics at Vienna, and thus Austria at least recognized Boltzmann's long battle for the recognition of atomic theory in physics, Mach could accept neither Boltzmann's physics nor his philosophy. This rejection may well have played a role in

Boltzmann's suicide in the spring of 1906. But by that time a younger man—at the Patent Office in Berne—had locked horns with these issues and decided in separate papers to follow the leadership of both Mach and Boltzmann (and Planck, too, for that matter) in assuming their doctrines for separate purposes.

Mach's historical account of the development of the science of matter in motion was consciously anthropocentric, beginning with ancient Greek contributions to the principles of statics. Then he considered both in text and in context the achievements of Galileo, Huygens, and Newton in establishing the principles of dynamics. By analyzing elucidations of mathematical mechanics through the eighteenth and nineteenth centuries, Mach was able to elaborate his severe criticism of the "metaphysical obscurities" inherited from the giants of the seventeenth century. Stressing his notions of the "economy of thought" and of the need for constant analyses of sensations, Mach thoroughly chastised Newton's postulates of absolute time, space, and motion, as physically meaningless:

> *All* masses and *all* velocities, and consequently *all* forces, are relative. There is no decision about relative and absolute which we can possibly meet, to which we are forced, or from which we can obtain any intellectual or other advantage. When quite modern authors let themselves be led astray by the Newtonian arguments which are derived from the bucket of water, to distinguish between relative and absolute motion, they do not reflect that the system of the world is only given *once* to us, and the Ptolemaic or Copernican view is *our* interpretation, but both are equally actual. Try to fix Newton's bucket and rotate the heaven of fixed stars and then prove the absence of certrifugal forces.[10]

Convinced of the artificiality, despite its utility, of the physicists' conceit of "isolated systems," Mach argued that "every single body of the universe stands in some definite relation with every other body in the universe." This notion, later christened Mach's principle, led the physicist to distrust the mathematician's confidence in his rational mechanics. Rest itself is always a case of motion in some larger or smaller context. Considering the formal development of analytical mechanics, Mach took a consciously anthropological and naturalistic stance to criticize the mythologies of mechanism and materialism. Mental fetishes of all sorts distort scientists' attitudes toward their methods, materials, and so-called laws:

> We must admit . . . that there is no result of science which in point of principle could not have been arrived at wholly without methods. But, as a matter of fact, within the short span of a human life and with man's limited powers of memory, any stock of knowledge worthy of the name is unattainable except by the *greatest* mental economy. Science itself, therefore, may be regarded as a minimal problem, consisting of the completest possible presentment of facts with the *least possible expenditure of thought.* [11]

Except for Mach's uncompromising anti-atomism and his sympathy for the thrust of the energetics movement, his teachings were so compelling for young physicists that many took him at his word, thus failing to see the philosophical inconsistencies into which they were led. Mach's relativity implied the functional interdependence of all sensations and their sources in the physical world. His epistemology, like most of those in the positivist tradition, was so skeptical ultimately as to verge on solipsism, idealism, or presentationalism. These problems led Lenin, for instance, to denounce Machist tendencies among Marxists in 1909, and eventually Einstein too would discover the patronage and position of his early hero, Ernst Mach, to be untenable.

Lorentz's Theoretical Physics

H. A. Lorentz was fifteen years younger than Ernst Mach, but having concentrated on theoretical physics rather than spreading himself broadly, he was more respected among his fellow specialists than was Mach. Being in the midst of his most productive period and having just shared the Nobel Prize for 1902 with his student Pieter Zeeman, Lorentz was widely known for his theory of electrons and the electromagnetic aether. He was the paragon theoretician attempting to fuse British (Maxwellian) and Continental (Weberian) electrodynamics into one coherent and consistent theory.

Helmholtz, Hertz, and Lorentz had carried through sequentially the main effort to match the insights of Maxwell with those of Weber and Clausius. Both the continuous and the discrete aspects of electricity thus were of primary concern to Lorentz. His formula for the force connecting the continuous field with discrete electric charges (which we first met here on p. 124),

$$E = 4\pi c^2 d + v \times H$$

had become, since 1895 when first offered in his famous *Treatise*, a major tool of the trade of physics.[12]

Meanwhile, however, Joseph Larmor at Cambridge University was the theoretician most nearly abreast of Lorentz. The two had wrestled independently for almost a decade with the implications of subatomic particles pervading continuous fields. Larmor's electromagnetic definition of mass and the FitzGerald-Lorentz contraction hypothesis had come into such repute in theoretical physics that no one interested in such problems could avoid their influence.

A young privatdocent at the University of Göttingen, Max Abraham (1875–1922), began developing a rival theory of electrons in 1902, which he hoped would supplant Lorentz's theory. By 1904, Abraham's theory of electricity, based on spherical nondeformable electrons, had the support of experimental evidence from Walther Kauffmann's studies of radium rays traveling through electromagnetic fields. But Lorentz persisted in his own development of the equations seemingly necessary to describe the behavior of streams of electrons traversing an electromagnetic field. Extremely conscientious toward old and new experimental evidence, Lorentz presented his transformation equations for second-order cases (v^2/c^2) in a paper entitled "Electromagnetic Phenomena in a System Moving with Any Velocity Less than That of Light" (1904). This paper began with Lorentz's old obsession with the problems pre-

Ernst Mach (*courtesy AIP Niels Bohr Library—Burndy Collection*)

Henri Poincaré (*courtesy AIP Niels Bohr Library*)

Max Planck *(courtesy AIP Niels Bohr Library)*

Albert Einstein *(courtesy the Einstein Estate)*

sented by the Michelson-Morley aether-drift experiment, and it ended with an attempt to accommodate Abraham's theory and Kaufmann's measurements. Lorentz's deductions proceeded from two assumptions: (1) that all electrons are ellipsoids flattened in the direction of their motion and (2) that the velocity of light is a speed limit for material particles, which leads to the supposition that motion of translation itself produces a deformation. Applying his dualistic transformation equations, which presumed a difference between general (true or absolute) time and local time and another difference between a continuous field (the aether) and discrete particles, Lorentz was able to codify a number of radical departures from classical mechanics in a form that separated field from matter theory and that unified relative mass and energy transfers for particles, electrons, and molecules in uniform rectilinear translations.[13]

Although Einstein claimed not to know of Lorentz's 1904 work until well after 1905, the mathematics and physical interpretations of these two theories were so similar as to lead to much confusion over priorities. The historical identity of the original notions for the advent of relativity theory is still debatable because of these similarities. One most obvious difference between Lorentz and Einstein, however, resides in the explicit scholarship of Lorentz and the implicit philosophy of Einstein. We shall see more of the latter later, but here we should notice what current concerns, besides Abraham and Kauffmann, were most relevant to Lorentz's work of 1904.

In 1900, Henri Poincaré had criticized all theoretical physicists for their tendency to introduce new ad hoc hypotheses, such as the FitzGerald-Lorentz contraction, in order to explain Michelson's null results for aether-drift. Lorentz was sensitive to Poincaré's objection, and so explicitly set forth to refine his earlier work, taking into account more recent experimental reports related to the relative motion of Earth and the aether of space. Meanwhile, Poincaré published the first edition of his influential book on epistemology, *Science and Hypothesis*, in 1902, and his praise for Lorentz, contrasted with his critique of Larmor, must have seemed encouraging to the Dutch theorist. Although the mathematician Poincaré acknowledged experiments as the sole sources of new truths and certainty about nature's laws, he also argued forcefully that men may be deceived as much by what they believe as by what they see.

A notorious case of self-deception in physics arose in France at the University of Nancy about this time. Professor René Blondlot claimed to be demonstrating a new species of "light," supposing it to fill in some gaps in the electromagnetic spectrum left by spectroscopists exploring around and beyond cathode, canal, and X rays. Blondlot's rays, christened "N rays" after the city and university that gave sustenance to their discoverer, were repeatedly reported in *Comptes Rendus* as periodic results of Blondlot's experiments to polarize X rays. Blondlot wrote fifteen papers and a book on the subject, exciting a wide and interested, if somewhat skeptical, audience. Then, unsuccessful attempts of other experimenters to repeat his work led to his retirement in disgrace. Robert W. Wood, an American physicist who had also suffered from skeptical superiors (as a graduate student under Michelson), was one visitor to the Nancy labs who reported, devastatingly, the differences between Blondlot's will-to-believe and his objective phenomena.[14]

If the case of Blondlot's nonexistent "N rays" exhibited the self-correcting inertial guidance that is supposed to control the advance of scientific disciplines, it also demonstrated how fitful and erratic was the progress being made during the progressive prewar era when the expansion of physical science in all directions at once was endemic. Experimental physics was expected to have growing pains in adjusting to its exciting new discoveries, to the commercial possibilities arising out of research in electrical and electronic phenomena, and to the need for proper training of the many new people interested in applied science, in military usefulness, and in quick returns from professional investments. However, few could expect that the current state of theoretical physics might not be based so solidly on empirical evidence as was supposed by many eminent leaders of science.

Such was the case, however, with the already widely celebrated aether-drift experiments. Poincaré, Larmor, and Lorentz had broadcast praise for the Michelson-Morley experiment as synonymous with the highest precision and the "nullest" result for an aether-wind test. Yet those experimentalists who were closest to the problem, knowing how tenuous had been their first efforts, decided to try again to trap nature into disclosing the relative motion of Earth against the aether.

In the decade and a half that had elapsed since Michelson and Morley had verified the trustworthiness of Fizeau's aether-drag experiment and then falsified the idea of a *stagnant* universal field of luminiferous aether, their reputation as iconoclasts had come to pervade the European community of physicists. Lorentz and Poincaré, Kelvin, Rayleigh and Planck, among others, had so often reinforced each other in praising the precision of Michelson-Morley that the incomplete experiment had acquired a reputation far in excess of its actual character. The awesome accuracy of Michelson's many other optical measurements and devices—especially the conversion of the length of the meter bar in Paris to a count of the wave lengths of a spectroscopic line of cadmium light—made him an authority whose experimental design and performance were seldom questioned. The unfinished aether-drift experiment of 1887 had obtained the status of a classic test largely on ancillary grounds (for instance, as a result of Michelson's 1895 vertical interferometer experiment) and because of theoretical interest rather than experimental repetition.

Both Michelson and Morley were somewhat ashamed of this unearned furor. Theoretical critiques by Maxwell's heir, William M. Hicks, and by Wilhelm Wien elicited a few experimental responses between 1902 and 1904. Michelson in Chicago reconsidered his old calculations along with some new proposals to beat the problem of synchronization; Morley and Miller in Cleveland did likewise with "The Theory of Experiments to Detect Aberrations of the Second Degree." Michelson, however, was too busy with a new family and various projects, particularly his ruling engine for diffraction gratings and echelon spectroscope, to return to aether-drift tests for the moment.[15]

Morley and Miller, however (as teaching duties would permit between 1902 and 1906), tried to perfect first a wooden-base, then a steel-base interferometer of the classic experimental design but with a path-length extended by almost a factor of ten. Their objectives were confused by the state of the theory, but they specifically set out to detect the hypothesized FitzGerald-Lorentz contraction, using different materials for the base and for the path-length gauges in their interferometers. After several years' effort their results turned out null again, and they reverted to the position that this was simply another, more refined version of the test for aether-drift on the classical model for measuring Earth's orbital motion.

In 1902, Lord Rayleigh, who like Kelvin and Lorentz had followed the aether-drift problem for over two decades, reasoned that transparent objects ought to show a second-order double-refraction effect if intermolecular forces were variable according to their orientation with respect to Earth's orbital motion. Rayleigh made some preliminary tests with water, carbon disulphide, and stacked glass plates but found no significant changes in their refractive properties regardless of how oriented or when observed. This judgment from so respected an authority on classical wave mechanics was likewise above suspicion. And yet shortly thereafter D. B. Brace of the University of Nebraska vastly improved Rayleigh's apparatus and techniques, trying in 1904 to find third-order evidence for the FitzGerald-Lorentz contraction. Brace, too, admitted defeat in the effort to find this delicate optical effect that would correspond to a hypothetical aether-drift. [16]

Although FitzGerald died in 1901, his associate F. T. Trouton, together with H. R. Noble of London, designed another, quite different type of experiment the next year "to discover, if possible, whether there be a relative motion between the earth and the aether." Setting up and charging an extremely sensitive and well-protected electric condenser, suspended like a torsion balance by a fiber in a vertical vacuum tube, Trouton and Noble sought a mechanical rather than an optical effect, yet also another second-order manifestation of an electromagnetic aether-drift. This suspended consenser should, in accord with astronomical calculations, reveal Earth's resultant motions if there were a stationary aether pervading all space. Without analyzing in much depth what astronomers were saying that total resultant motion should be, Trouton and Noble concluded in 1903 that there could be "no doubt that the result is a purely negative one." [17]

Thus, by alternative experiments and repeated failures to confirm an aether-wind, the news from the experimental side of the house of physics began to persuade the theoretical side that aether-drift tests were hopeless. Historically, it seems fair to say that by 1904 these varied experiments had produced enough null evidence to warrant the assumption that Earth's relative motion is not measurable without reference to something more substantial than the luminiferous or the electromagnetic aether.

Although Lorentz was not fully aware of all these developments, he knew and cited most of them in the paper presenting his second-order transformation equations of 1904. It is hard to believe that Einstein was not equally aware of Lorentz's more recent work. But we know he was studying Poincaré's *Science and Hypothesis*.

Poincaré's Principle of Relativity

Since at least 1900, Poincaré had been speaking in various contexts of a "principle of relativity" as an empirical recognition of the rule that science can aim to know only the relations between things and never things in themselves. In his widely read book of 1902, *Science and Hypothesis*, he endeavored to instill a proper balance of pride and humility in understanding the achievements and limits of scientific thinking. Like Mach in some respects, especially regarding the anthropomorphic origins of mechanics, Poincaré was an aesthete who defended creative hypotheses and stressed conventions rather than economy of thought. Considering the relativity of space introduced by the new geometries and the role of experiment in testing mathematical conventions, Poincaré talked of the "law of relativity," almost as Maxwell had spoken of a "doctrine," with respect to phenomena of state and position measurements connecting experiments with geometry.[18]

His seventh chapter, entitled "Relative and Absolute Motion," was built entirely around the "principle of relative motion." From this exposition, any reader like young Einstein, raised to believe in the alleged certainty of mathematics, especially geometry as applied to mechanics, could only be profoundly shaken in that belief. For Poincaré's discussion of the role of hypotheses in physics was a compelling critique of the metaphysics to which both experimentalists and mathematicians often held unconsciously. Sparing none (except Mach) of the great theorists—Helmholtz, Maxwell, Hertz, Kelvin, Larmor, Lorentz—Poincaré discussed at length, almost as if conversing with them, the unsolved problems bequeathed by the masters of physics. He emphasized a variety of topics: the non-Euclidean geometries of Lobachewski and Riemann, the "world of four dimensions," classical mechanics, relative motion, and then the theories of electromagnetism. The concept of the electromagnetic aether served him well as an illustration of the quest for continuity and unity in physics and as a symbol of the metaphysics that so many physicists tacitly accepted without analysis as part of their inheritance.

Poincaré provided a classic statement of the paradoxical march of science toward antithetical goals: unity and simplicity on the one hand, and diversity and complication on the other. Which tendency would prevail was an issue for faith and the future to decide. But science itself would be impossible without a willingness to act as if certain conventional propositions were true. The drive toward unity and simplicity therefore should be encouraged, although always it should be tempered by self-criticism.

Repeatedly, Poincaré asked whether the aether actually exists. Knowing the origin of the idea of a luminiferous aether, that light takes time to travel through space and that during that time its waves must be extant somewhere between transmission and reception, Poincaré nevertheless argued that the electromagnetic aether was a pragmatic convention, merely a useful concept so far without empirical support. After discussing the delicate questions raised by the calculus of probabilities applied to physics, he turned in the last two chapters in *Science and Hypothesis* to consider "optics and electricity" and, finally, "electro-dynamics." Praising optics, especially the undulatory theory, as the most advanced branch of physics, Poincaré wrote:

Whether the ether exists or not matters little—let us leave that to the metaphysicians; what is essential for us is, that everything happens as if it existed, and that this hypothesis is found to be suitable for the explanation of phenomena. After all, have we any other reason for believing in the existence of material objects? That, too, is only a convenient hypothesis;

only, it will never cease to be so, while some day, no doubt, the ether will be thrown aside as useless.[19]

Having praised the form of Maxwell's synthesis of electricity, magnetism, and optics, Poincaré went on to criticize the *Treatise* of 1873 as extremely ambiguous in its substance. Noting differences in national styles between French and English approaches to electromagnetism, he nonetheless felt that Maxwell should have made up his mind more definitively. Yet Poincaré recommended that physicists confine themselves to questions that are accessible to positive methods, and he praised the dual achievements of Maxwell's theory and Hertz's experiments as marking a marvelous new synthesis well beyond the works of Ampère, Weber, Helmholtz, and others. Rowland's experiments on electromagnetic "convection currents" and Lorentz's theory of electrons appeared to Poincaré quite attractive in 1901 as he was preparing *Science and Hypothesis* for the press. But he closed his book with certain objections that were an open invitation for someone like Einstein to catch and redress in more comprehensive form:

The phenomena of an electric system seemed to depend on the absolute velocity of translation of the centre of gravity of this system, which is contrary to the idea that we have of the relativity of space. . . . Imagine two charged conductors with the same velocity of translation. They are relatively at rest. However, each of them being equivalent to a current of convection, they ought to attract one another, and by measuring this attraction, we could measure their absolute velocity. "No!" replied the partisans of Lorentz. "What we could measure in that way is not their absolute velocity, but their relative velocity *with respect to the ether*, so that the principle of relativity is safe." Whatever there may be in these objections, the edifice of electro-dynamics seemed, at any rate in its broad outline, definitively constructed. Everything was presented under the most satisfactory aspect. . . . Numerous investigators have endeavored to solve the question [of the electrodynamics of moving bodies], and fresh experiments have been undertaken. What result will they give? I shall take care not to risk a prophecy which might be falsified between the day this book is ready for the press and the day on which it is placed before the public.[20]

Einstein, Besso, Solovine, and the Habicht brothers, together with other visitors to their "Olympian" circle in Berne, read and discussed Poincaré's little book with great relish. Their attitudes toward the whole scientific enterprise were here reinforced and given direction. But meanwhile Poincaré himself proceeded much further.

In 1904, Poincaré and Paul Langevin were among the savants invited to attend the World's Fair in St. Louis, Missouri. Here, at another International Congress of the Arts and Sciences, Poincaré expressed himself even more pointedly on "The Principles of Mathematical Physics." Having earlier compared the house of physics to a library wherein experimentalists acted as acquisitions people and mathematicians as the cataloging staff, Poincaré now spoke of the present crisis in physics as brought on by more significant data than had as yet been classified and assimilated into a coherent and usable system. Two of the six or so foremost physical principles that seemed in jeopardy were those mechanical relations symbolized by Newton's third law of dynamics, the action-reaction equilibria, and by Galileo's principle of relativity as applied to the electrodynamics of moving bodies. Critical of Lorentz once again and with highest praise for Michelson's aether-drag and aether-drift experiments, Poincaré spoke of electrons and the aether, the problem of their mass and energy relationships, and the need for theoretical clarification of the most basic notions in physics. To save the principle of relativity and to explain the variability of electrodynamic masses, Poincaré specifically suggested that the velocity of light might be taken as a new absolute constant:

From all these results, if they are confirmed, would arise an entirely new mechanics, which would be, above all, characterized by this fact, that no velocity could surpass that of light, anymore than any temperature could fall below the zero absolute, because bodies would oppose an increasing inertia to the causes, which would tend to accelerate [negatively] their motion; and this inertia would become infinite when one approached the velocity of light.[21]

Whether or not young Einstein studied this and other such pronouncements by Poincaré in the periodical literature of 1905, the master mathematician of France proceeded to amplify his thoughts on the "postulate of relativity" during that year. Based on Michelson's null results for aether-drift and the FitzGerald-Lorentz contraction hypothesis, Poincaré wrote a lengthy paper "on the dynamics of the electron," which defined "relativity" in terms of the evidently impossible effort to determine experimentally the absolute motion of Earth. This important technical study was completed in July 1905 but was not fully published until the next year and then only in an Italian mathematical journal not regularly read by European physicists. By virtue of its contents, however—dealing directly with the Lorentzian concerns over invariance and covariance in transformation equations, with the theory of the electron, and with hypotheses relating to electromagnetic and gravitational forces—Poincaré's so-called Rendiconti paper on relativity demonstrates clearly how the main ideas for which Einstein later became famous were also being identified elsewhere in the ambient atmosphere of the times.[22]

And yet because of a fourth professional influence on Einstein, that of the leading mathematical physicist in Berlin, namely, Max Planck, the combination of trends that the Swiss patent examiner was to personify grew into a unique and exquisite form. Planck was soon to replace Drude as editor-in-chief of *Annalen der Physik*, and this replacement in 1906 symbolizes the shift in German physics from the predominance of the paradigm of continuity to that of the paradigm of discreteness.

Planck's Quantum and Hasenöhrl's Claim

Max Planck was twenty-one years older than Albert Einstein, and being the foremost German theoretical physicist at the time Einstein was maturing, as well as associate editor of *Annalen der Physik*, he naturally became a model for the younger man to emulate. There was much in Planck to admire, but Einstein's independence was such that his judgment of Planck's achievements was almost as important as Planck's judgment of Einstein. The two became in time complementary and reciprocal in certain ways, but that relationship grew slowly through several decades. Most significantly, however, Planck's theoretical derivation of the quantum of action in 1900 received its first major recognition from Einstein's 1905 paper on the photoelectric effect, which suggested the heuristic idea of light quanta. Because Planck himself vacillated between the aether and quantum hypotheses and between his interests in thermodynamics and electrodynamics, Einstein's reinforcement of Planck's quantum theory came at a most opportune time. It tended to ignite the professional development of quantum theory, and it probably was indeed, as Einstein suspected, the most revolutionary of his three main papers of the year 1905.[23]

Most accounts of the advent of twentieth-century physics obscure the historical situation in the first decade or so because of the need for clarity in presenting the concurrent advent of the quantum theory. The need to show the growth of confidence in the new physical atomism seems to override the need for honesty in presenting the confused and conflicting turmoil of the times. Planck himself, however, often confessed how tumultuous was his own struggle with old notions and new ideas regarding light, heat, energy, and entropy. His Nobel Prize address of 1920 began with an allusion to Goethe's saying that man errs as long as he strives, and it ended with the dilemma, then still rampant, raised by the confrontation of classical wave theory with the neoclassical particle theory.

Between October and December 1900, while working on the problem of the energy distribution emitted from a Hertzian oscillator and absorbed by its resonator, Planck had experienced his first major insight into the quantum of action. Not yet trusting Lorentz's electron theory and being thoroughly imbued with Maxwell's, Kirchhoff's, and Wien's approaches toward energy exchange, Planck found a way to simplify the calculations involved in studying the connection between energy and temperature. Three famous "laws" were associated with each name, respectively: that the electromagnetic spectrum is continuous, that all hot bodies regardless of their chemical nature emit about the same color balance at the same temperature, and that their energy distribution curves shift uniformly with changing temperature. These laws needed integration and seemed basic enough to convince Planck that he was near the discovery of "something absolute." From an empirical emission formula that quickly proved experimentally successful, Planck derived encouragement, and he plunged intensively into the search for a hypothesis that would unify Kirchhoff's law and Wien's displacement law.

Wrestling with calculations of energy and temperature and then of energy and time as related to the connection between probability and entropy, Planck found that he had to make a radical change in classical assumptions regarding the emission of energy. The observed facts seemed to demand that any single submicroscopic electric oscillator within Hertzian oscillators or J. J. Thomson's vacuum tubes worked only in spurts, radiating energy only in integral multiples of the frequency of its vibrations and some universal constant. Extended to collections of all the uncountable oscillators on the surface of a real glowing body, Planck's quantum hypothesis accounted for the continuous spectrum, for the temperature–color balance relation, and for Wien's energy-distribution law.

Much later this extraordinarily simple new relationship was further simplified into the formula

$$E = h\upsilon$$

where h is Planck's quantum of action or constant and υ is the frequency of oscillation. Planck's first calculation of the product of energy and time gave a value of 6.55×10^{-27} erg-sec, and this value has stood the test of time very well, being refined only to 6.52×10^{-27} erg-sec by 1920 (today it is usually given as 6.626×10^{-34} joule-sec). Although Planck's radical assumptions of energy levels and discontinuous quanta for radiation were direct assaults on habits of thought inherited with the infinitesimal calculus, wave theory, and field theory, the idea of energy quanta gradually proved so valuable and Planck's constant proved so fundamental that the new theory began to pervade all of microphysics. The basic continuity of all causative connections was challenged by this equally basic discontinuous view of the world.[24]

Yet Planck himself was loath to admit this until long after Einstein offered his support in 1905 by arguing for a heuristic acceptance of light quanta in *propagation* as well as *emission* processes. Hertz's photoelectric effect, Owen W. Richardson's law for thermionics, and Rutherford's studies of radioactivity, isotopes, and transmutations of matter and energy all intervened to complicate the picture. Planck meanwhile held fast to the paradoxes he had introduced—the beginnings of the photon-wave dilemma—by denying that one could imagine a physical picture of what was happening (as proved by the observed numbers) at so fundamental a level of interactions. And he also came to defend the physicists' faith in a sophisticated realism, that the physical world exists apart from man's imperfect sensing of it.[25]

That Einstein also worried over such problems is well known, but that another young physicist in Vienna achieved similar solutions to similar worries is now almost forgotten because of later prejudices and circumstances. Friedrich Hasenöhrl (1874–1915), who came from an aristocratic Austrian family, was a student of Stefan's and Boltzmann's, and a teacher at Vienna's Institute of Technology. In 1904, he published (before Einstein) a paper in *Annalen der Physik* that suggested a form of electromagnetic mass-energy equivalence related by the square of the speed of light. This, together with a clever thought-experiment modeled on Kirchhoff's idea for a hollow "black-body" cavity resonator, led to such recognition that Hasenöhrl succeeded Boltzmann in the chair of physics at the University of Vienna after the latter's suicide in 1906. At the outbreak of World War I in 1914, Hasenöhrl volunteered for military duty, was wounded in action, and finally was killed at the Battle of Vielgereuth. Hasenöhrl's rivalry with Einstein may have been uneven, but there can be no doubt that nationalistic passions and ethnic bigotries have affected evaluations of their respective historical roles. Philipp Lenard's "Aryan" hero was Hasenöhrl.[26]

By 1904, therefore, renewed awareness of an impasse in experimental attempts to detect an aether combined with real advances in electronic experiments to inspire theoreticians to try new departures. Despite the demise of confidence in the aether and the rise of the field concept as a replacement, there were few critics who, like the French philosopher Henri Bergson, attacked physics for its dichotomy between spatial and temporal images of the inanimate world. Owing largely to the influence of modern astronomy, spatial consciousness led directly to an increased temporal consciousness: to look into an astronomical telescope was to look backward incredibly far in time. Likewise in atomic physics to probe in imagination disciplined by experiments the events of particle interactions was to grope for meaning regarding sequential and specific order. Since no single molecule or atom, much less an electron, could be stopped, isolated, and studied as distinctly unique, why talk about them with definite adjectives? What could our gross measures of space and time mean in the microcosmos and macrocosmos?

Aside from Mach, Lorentz, Poincaré, and Planck, who were expected to be on the verge of offering a radical departure, some younger contenders also seemed likely to cut such a Gordian knot. Other than Hasenöhrl, the leading candidates probably were Joseph Larmor, Paul Langevin, Walther Ritz, Wilhelm Wien, Marcel Brillouin, Max Abraham, and Hermann Minkowski. But the seed came not from famous names or students of famous people; the decisive insight came out of a discussion group of Swiss-trained "graduated professional students" who scattered around Europe and gathered in Berne occasionally to talk about the constancy of the speed of light, among other things. Their first among equals was Albert Einstein, patent official (third class), amateur violinist, frustrated academic, and philosopher of nature.

Einstein's Postulates and Minkowski's Patronage

In retrospect, it seems that science after 1900 was awaiting the appearance of a physical theory that would give the concept of time equal billing with the dimensions of space. "Relativity" in general, deriving much from the relativity of ethical values, culture, knowledge, and history, had been very much in the air of the late nineteenth century. Positional astronomy through photography and photometry had developed ever more sophisticated means of measurement and relative systems of reference. By 1905, J. C. Kapteyn, a Dutch astronomer with a profound interest in perfecting William Herschel's value for the solar apex, had identified two star streams within our Milky Way galaxy, thus further complicating the absolute-motion problem for our solar system. Mathematics had progressed well beyond non-Euclidean geometries and linear algebras into an age of analyses of more curious kinds of convoluted relationships. "Reality" was widely admitted to be whatever the mind could predict and mental or sensory experience could corroborate.[27]

Mechanics almost alone was hobbled by a virtually finished world-view. Whoever could crack this confidence was practically assured of a notoriety or a notability without parallel. In particular, Poincaré's partial solution to the n-body problem (relating to the interacting attractions of all bodies in the solar system and beyond) had led to his pronouncement of the need for extending the principle of relativity. After Rutherford and Frederick Soddy (a young chemist at McGill University in Montreal), in 1902, showed how radioactive "emanations" could result in the transmutation of matter, a new link between energy and matter was suggested. If anyone could weld mechanics with electromagnetism by means of mathematically forged ties between mass and energy, then time might finally get its just consideration as covalent with space when dealing with extreme experiences.

In the effort to understand the roles played by extremely long or short times in physical phenomena, a fundamental difference arose between the seasoned H. A. Lorentz and the neophyte Albert Einstein. Lorentz relied on local times, theoretically construed, for his ultraminute electrical charges at work in fields of space. These he assumed were synchronizable with other particles at work in other fields. Lorentz apparently felt little need until well after 1905 to analyze what it would mean operationally to try to synchronize clocks referrable to tiny particles moving near the velocity of light. Einstein tackled this problem after he became aesthetically disgruntled with the asymmetries provoked in theory by the Maxwell-Hertz-Lorentz tradition of analysis of the electrodynamics of moving bodies.

In 1901 and 1902, Lorentz was lecturing often on aether theories and aether models. He saw the problem of the continua in four parts: analyzing first the aberration of starlight as previously studied by Fresnel, Stokes, and Planck; then going on to mechanical aether and elasticity theories, particularly those of Maxwell and MacCullagh; next looking at Kelvin's model of the aether from the standpoint of vortex atoms; and finally considering abstractly the attraction and repulsion of pulsating spheres. He treated the thermodynamics, entropy, radiation, and quantum aspects of these problems at length. The Michelson-Morley experiment bore a special place in the illustration of his thoughts, as was usual with Lorentz.[28]

Prior experience with five papers in Drude's *Annalen* from 1901 through 1904 had set Einstein's pattern for the publication of his three most insightful articles of 1905. The physical focus for all three papers was a profound concern for fluctuations in radiation pressure. Central to the last, most famous one, "On the Electrodynamics of Moving Bodies," was an analysis of simultaneity, relative motion, and the measurable meaning of space and time near their conceivable limits, both cosmic and atomic.

The first paragraph of Einstein's most famous paper drew attention to the asymmetry in Maxwell's electrodynamics whereby magnets and conductors in relative motion were treated differently depending on which was considered to be at rest. Because only their relative motion is inherent in the production of the electric current, the physical equations seemed to him obviously more complicated than nature itself. And because all attempts to discover the absolute motion of Earth relative to a "light aether" had failed, Einstein suggested that the whole idea of absolute rest be discarded. Rather than follow Lorentz, he proposed to extend Galilean relativity for the laws of mechanics to cover the laws of electrodynamics and optics as well:

We will raise this conjecture (the purport of which will hereafter be called the "Principle of Relativity") to the status of a postulate, and also introduce another postulate, which is only apparently irreconcilable with the former, namely, that light is always propagated in empty space with a definite velocity c

which is independent of the state of motion of the emitting body. These two postulates suffice for the attainment of a simple and consistent theory of the electrodynamics of moving bodies based on Maxwell's theory for stationary bodies.[29]

With this statement of his two principles raised to the status of postulates, Einstein proceeded to discard also the luminiferous aether as a superfluous notion. The kinematical part of his paper then launched his epistemological discussion of rigid bodies (equivalent to coordinate systems), clocks, and electromagnetic processes. And here he redefined the notion of simultaneity as the primary operational problem, leading to the relativity of length and time in extreme cases of transformation or translatory or uniform rectilinear motion.

Einstein, from Mach's viewpoint, saw Lorentz's problem (from the old formulation of 1895) as tied directly to the assumption that time flows on regularly, linearly, irreversibly, and forever without regard to physical events that happen to give us benchmarks. This Lorentzian reliance on *allgemeine Zeit*, general or absolute time, had its parallel in the Newtonian assumption of an absolutely stationary space, which also seemed to Lorentzians a necessary philosophical presumption in order to allow physics to operate with reliance on *Ortzeit*, local time. Like those conventions adopted by mariners and railroad men to make sense of daily life dominated by clocks on long journeys eastward or westward, local time zones could be adjusted to fit the longitude; they could be compared to match some reference meridian and standard clock. Though new to traveling cosmopolites and only recently sanctified by international treaties, local times had long been familiar to astronomers and mariners. There seemed no reason to Lorentz why he could not adapt this traveler's convention to the electronic world.

To Einstein, however, the reason one could not use local times legitimately was obvious: if the finite speed of light is constant and yet "plays the part, physically, of an infinitely great velocity," then one "instant" on any particle moving very near that speed limit could not be synchronized on any other particle or body; nor could there be any absolute reference frame, any natural prime meridian or GMT (Greenwich Mean Time) by which to compare local instants. Observers of "instants" on particles, planets, or stars could not exchange the information necessary to check whether any hypothetical clocks (or measuring rods) in two or more such systems were precisely aligned. To Einstein, there simply was no need for *allgemeine Zeit* at all; absolute time, like absolute space or absolute motion or the hypothetical stationary aether of space itself, was *überflüssig*, quite superfluous. The *inherent ambiguity of instantaneity*, in short, *precludes any ability to establish absolute simultaneity*.

Here, where the concept of simultaneity applied to kinematics, Einstein found his most subtle and telling point: if there cannot be any *absolute* significance to the concept of simultaneity, then Lorentz's general time and space and all other assumptions of *absolute* motion must be physically, that is, electrodynamically, unreal or meaningless. Because the principle of relativity could be made compatible with the postulate of the constancy of the velocity of light, it followed that the Maxwell-Hertz-Lorentz equations for empty space could be reduced to the assertion that "electric and magnetic forces do not exist independently of the state of motion of the system of coordinates." Physically this led Einstein to his key insight:

What is essential is, that the electric and magnetic force of the light which is influenced by a moving body, be transformed into a system of coordinates at rest relatively to the body. By this means all problems in the optics of moving bodies will be reduced to a series of problems in the optics of stationary bodies.[30]

Chronometers must behave differently at Earth's equator and at its poles, with slower time showing wherever the resultant velocity of compounded movements is greater. Also, electrons should demonstrate the symmetry of the new theoretical equations by rendering possible new measures of velocity, potential difference, and curved paths in magnetic fields. Thus the masses of particles and planets depend upon their velocities. Having shown that the concept of simultaneity is relative to the coordinate system chosen for its measurement, Einstein went on to show that the Galilean transformation equations were inadequate. He derived the same transformation equations from the Maxwell-Hertz equations that Lorentz had lately constructed. He then applied them to the dynamics of nonradiating electrons, offering three possible experimental suggestions to test his ideas.

Toward the end of the electrodynamical part of his main paper of 1905, Einstein briefly alluded to the implications of kinetic and potential energy differences for the concepts of longitudinal and transverse mass as applied to slowly accelerated electrons. Although not yet explored, here was the seed for an equation, later changed in notation and simplified as

$$E = mc^2$$

which expressed the first and perhaps most significant of the unifying insights that Einstein gave to physics. The idea of the equivalence of mass and energy symbolized by this expression came to be universally recognized, if not fully appreciated, by midcentury.

In 1905, almost as an afterthought, Einstein published another brief note entitled with a question, "Does the Inertia of a Body Depend upon Its Energy Content?" This momentous appendix to the electrodynamics paper began the theoretical development that was to guide much experimental work over subsequent decades in demonstrating the mass equivalence of energy. At the beginning, however, Einstein merely concluded that "radiation conveys inertia between emitting and absorbing bodies." He firmly asserted that "the mass of a body is a measure of its energy-content."[31]

When Einstein had finished his first publications on what he then preferred to call his "theory of invariance," there were very few outside his intimate circle applauding for several years. At first hardly anyone read, understood, or tried to fathom what this unknown young writer without a laboratory or academic connections was trying to say. But Einstein continued to elaborate on his denial of absolute motion in principle and on his obsession with the operational meaning of the constancy of the speed of light. In 1907, his most prolific year thus far, Einstein published seven papers, using "Relativity Principle" in titles of three of them for the first times.

Despite the weight of experiments by Walther Kaufmann apparently falsifying Einstein's predictions, the young theoretician maintained his confidence that field theory could now be divorced from the aether and from the concept of matter. Whether or not his "heuristic" light quantum (eventually christened the "photon") should become the basis for a renewal of the corpuscular theory of light, the appearances of energy and mass would be saved. In 1907, Walther Ritz, prompted largely by the Michelson-Morley and other aether-drift failures, did try to re-establish the "emission" (or ballistic or corpuscular) theory of light. Einstein returned to his postulate of relativity to draw out the implications of his idea of mass-energy equivalence. In this he was encouraged by his old friend Marcel Grossmann, by Max von Laue, who visited Einstein personally to express his admiration, and by praise from a few highly placed figures such as Planck in Berlin and Otto Lummer and Ernst Pringsheim in Breslau.

In September 1908, Hermann Minkowski, who had been Einstein's occasional mathematical mentor at Zurich, gave his former student an important boost by paying him the highest compliment: crediting him before fellow professionals with an admirable idea. Now a distinguished young professor at Göttingen, Minkowski delivered a seminal address before the Eightieth Assembly of German Natural Scientists and Physicians meeting in Cologne. This talk, entitled *"Raum und Zeit"* ("Space and Time"), was expanded into a booklet and published shortly thereafter, despite the author's untimely death. Minkowski's long-standing concerns with quadratic forms and differential equations appropriate to relative-motion problems in electromagnetic field theory led him to conceive the four-dimensional manifold in which the idea of an inseparable hyphenated "space-time" was emphatically stated: "Henceforth space by itself, and time by itself, are doomed to fade away into mere shadows, and only a kind of union of the two will preserve an independent reality."[32] Although Minkowski's own formulation of this continuum did not take hold, his enthusiastic endorsement and advocacy of Einstein's work spread Einstein's name well across disciplinary boundaries and sped his fame into the topmost professional scientific councils. Minkowski's address also spurred Einstein himself to develop his synthesis of the general theory of relativity.

The Reception of Restricted Relativity

Most mathematicians and philosophers, conditioned by Min-kowski's influence, imagine the advent of the special theory of relativity primarily in terms of Einstein's particular critique of the notions of space and time. Most physicists imagine it in Einstein's efforts to preserve the invariance of the Maxwell-Hertz theory and of Lorentz's transformation equations for electronics. Einstein himself, though always deeply interested in the experimental problems of thermo- and electrodynamics, discovered in 1905 that he was primarily a theoretical physicist. He preferred thereafter to stress kinematics, the study of abstract mathematical motion, as applied to the dynamics of heat and radiant-energy transfer. For him the recognition of the mass-energy equivalence that resulted from his analyses of spatial and temporal measurements on extreme scales was one most satisfying aspect of the next several years.

However, Einstein's doctoral dissertation on a new method of determining molecular dimensions and his interests in Brownian motion and Planck's theory of radiation continued to be equally compelling as he strove to publish his work. Light emission, prop-agation and absorption, specific heat theory, and in general all connections between particles and fields captured his most intense interest.[33]

Einstein's quest for ways to unify physics, which by the beginning of 1906 had already produced the mass-energy linkage, resulted soon thereafter in his discovering another principle of equivalence, that between inertial and gravitational mass. While Minkowski and a very few other theorists and mathematicians were studying the space-time implications of the relativity principle, Einstein was preparing a long article, his first for a publication other than the *Annalen der Physik*, for the German *Yearbook of Radioactivity*. Here, in the volume for 1907, appeared the first explicit statements of the famous $E = mc^2$ equation and of the inertia-gravity linkage. This discovery—that gravitational fields like electromagnetic fields have only relative existences—Einstein later called the "happiest thought of my life."[34]

Thus in overview, Einstein's review paper of 1907 represents the nexus of his personal development from the restricted to the general theory of relativity. Although he continued for several more years to work over the implications of all three of his seminal papers of 1905, not until 1911 did he begin publishing more on gravitation. Only then did his first title specifically on "Relativity Theory" appear. But the overlay of four years between 1907 and 1911 was not only a period of gestation for his new principle of equivalence for inertial and gravitational mass; it was also a time of intense concern over the implications of quantum theory for both matter and radiation. In 1909, Einstein delivered in person his first major address before a meeting of professional scientists at Salzburg. There he argued that neither wave theory nor particle theory alone could suffice to explain radiation; rather, a "kind of fusion of the wave and emission theories" would be necessary for the future of physics. This preview of the wave-particle duality to come a decade or more later indicated once again the physical intuition, the instinctive quest for unity, and the polar paradoxes that Einstein was beginning to embody.[35]

Thanks mostly to his gaining his doctorate in 1905, Einstein was promoted to a "second class" expert in the Swiss Patent Office in 1906. Also that year he was flattered to be sought out by a young German physicist named Max Laue, Planck's assistant, who was visiting Berne. The next year, while Einstein was thinking his happiest thought, another young man, Jakob J. Laub, joined the "Olympian Academy" for a while and became Einstein's first collaborator on several papers. Also in 1908, without moving, Einstein was allowed to take his first step into academic ranks by offering courses as a privatdocent (a kind of apprentice teacher) at the University of Berne. Then, in 1909, he was called to Zurich's cantonal university, where he had earned his doctorate, to become associate professor of theoretical physics. Minkowski's patronage and the extraordinary sacrifice of that position by an idealistic friend, Friedrich Adler, helped assure Einstein of this recognition.[36]

Meanwhile, although the three main unifications of Einstein's lifework had already been adumbrated—mass-energy, space-time, gravity-inertia—the analyses of the principle of relativity, the "invariance theory," and its gravitational implications were still in embryonic states. Einstein was achieving recognition in modest ways as a theoretical physicist of varied talents, not simply as the author of a single theory.

In July 1910, in Zurich, the 35-year-old Mileva gave birth to another son, the ill-fated Eduard. Hans Albert was now six, and the household was lively, but the son Eduard suffered more than his father had as a child.

Einstein began to correspond more broadly with his fellow professionals, especially with Laub and Laue, two of the first enthusiasts for relativity theory. After the 1909 Salzburg meeting, Einstein began to impress a growing circle of academic colleagues, including Planck, Rubens, Wien, Sommerfeld, and young Max Born. But his self-assurance also rubbed raw sometimes. In general the older generation of physicists was still too deeply committed to the aether or to the electromagnetic world-view to take Einstein seriously. Although Planck and Mach seemed separately to be growing receptive to Einstein's relativity, Lorentz and Poincaré were more reluctant, it seemed, to acknowledge the value of raising the principle into a postulate.

Lorentz began speaking about Einstein's critique of his assumptions in public lectures as early as 1906, but the idea of the relativity of simultaneity seemed so radical, so threatening to his theory of electrons and the aether, that he waited several years before committing to print his rebuttal. In 1909, after reviewing his longstanding concern over the Michelson-Morley experiment and its more recent repetitions by Morley and Miller, Lorentz reaffirmed the contraction hypothesis and expressed his faith in the eventual reconciliation of Einstein's theory and his own:

> Einstein simply postulates what we have deduced. . . . By doing so he may certainly take credit for making us see in the negative results of experiments like those of Michelson, Rayleigh and Brace, not a fortuitous compensation of opposing effects, but the manifestation of a general and fundamental principle.
>
> Yet, I think, something may also be claimed in favour of the form in which I have presented the theory. I cannot but regard the ether . . . as endowed with a certain degree of sub-

stantiality, however different it may be from ordinary matter.[37]

In 1910 and 1912, Lorentz lectured on "The Principle of Relativity for Uniform Translations" and suggested that Einstein had really not abolished the aether so much as he had circumvented it. Lorentz was acutely aware of the experimental investigations on the mass, velocity, and shapes of electrons by Kauffmann, Bestelmeyer, Bucherer, and others. He felt uneasy with the philosophical import of Einstein's 1905 paper, remarking that all epistemological problems ought perhaps to be handed over to the philosophers altogether:

Whether there is an ether or not, electromagnetic fields certainly exist, and so does the energy of the electrical oscillations. If we do not like the name of "ether," we must use another word as a peg to hang all these things upon. It is not certain whether "space" can be so extended as to take care not only of the geometrical properties but also of the electrical ones.

One cannot deny to the bearer of these properties a certain substantiality, and if so, then one may, in all modesty, call true time the time measured by clocks which are fixed in this medium and consider simultaneity as a primary concept.[38]

By 1910, Lorentz believed there was a hung jury: the experiments of Kauffmann investigating the motion of free electrons had given evidence against the Einsteinian postulate of relativity and for Max Abraham's theory of the spherical electron. Bucherer's similar experiments of 1909 were accepted by Lorentz as evidence for the postulate of relativity and for his own competitive idea of ellipsoidal electrons. This experimental contradiction remained a problem well into the next decade, until Carl Neumann in 1914 and Charles-Eugène Guye with Charles Lavanchy in 1916 at long last announced careful confirmation of Bucherer's results. Thus, Lorentz's theory of flattened or contracted electrons came to supersede Abraham's theory, to vindicate the FitzGerald-Lorentz contraction hypothesis, and to support Einstein's relativity postulate. Thereafter, Lorentz could accommodate his views to Einstein's gladly and with honor.[39]

Poincaré remained strangely silent about relativity in public after 1906, however much or little he may have pondered Einstein's insights and derivations privately. By 1910, it was obvious that the raising of the principle into the postulate of relativity in Einstein's 1905 paper was a stratagem that could have few, if any, immediate effects in explaining experiments then possible in physics laboratories. Because the 1905 article neglected gravity and dealt only with "uniform rectilinear motion," a condition almost as rare in macro-nature as that of "absolute motion," Poincaré may have simply ignored the seriousness of Einstein's work. It had its applications only in micro-nature or particle physics where gravity is too weak to be taken into account. Poincaré was perhaps too much the celestial mechanician to worry over the restricted relativity hypothesis from young Einstein.[40]

The gravity of gravity, in other words the serious need for a more general theory of *accelerated* motions of bodies affected by one or more gravitational fields, could not be overestimated. Einstein might well have been forgotten had he not gone on to produce (as he began to promise about 1907) a more generally applicable theory of relativity.

Thus, only around 1911, as Einstein revealed hints and promises of more fertile and meaningful work to come, did his insights into relativity start to be called a "theory." Max Laue published the first book on the subject that year. Other analyses of relativity had been growing in number more recently, and they were often devastatingly critical, primarily because of Einstein's discard of the aether and the absolutes. But only after a *general* theory of relativity became expected imminently did the earlier *restricted* theory become widely known. Meanwhile, its author continued to contribute to physics more broadly with papers on wave mechanics, quantum theory, and fluctuation pressures. He was achieving a reputation in German science that was secure and respectable, whether or not his name was associated with relativity. In the course of another decade, the work of the years during and after 1905 on the problems posed by the postulates of relativity and the constancy of the speed of light would become celebrated as Einstein's special (or restricted) theory of relativity.[41]

NOTES

1. One of the best surveys of intellectual life leading into our century is still John T. Merz, *A History of European Thought in the Nineteenth Century*, 4 vols. (Edinburgh: Blackwood, 1896–1914), 4th ed.; see esp. Vol. II, p. 89, and Vol. III, p. 544. See also Fritz K. Ringer, *The Decline of the German Mandarins: The German Academic Community, 1890–1933* (Cambridge: Harvard Univ. Press, 1969); and Stanley L. Jaki, *The Relevance of Physics* (Chicago: Univ. of Chicago Press, 1967), pp. 52–94, 236–80, 330–70. For the state of the profession, see Paul Forman, John L. Heilbron, and Spencer Weart, *Physics circa 1900: Personnel, Funding and Productivity*, Vol. V of *Historical Studies in the Physical Sciences*, R. McCormmach, ed. (Princeton: Princeton Univ. Press, 1975).

2. The most trustworthy biography for this period is Carl Seelig, *Albert Einstein: A Documentary Biography*, trans. M. Savill (London: Staples, 1956), orig. Zurich, 1954. But see also the dual articles on Einstein by Martin J. Klein and Nandor L. Balazs in *DSB*, vol. IV, pp. 312–33.

3. For the generational conflict thesis, see Lewis S. Feuer, *Einstein and the Generations of Science* (New York: Basic Books, 1974), pp. 3–105. For good illustrations and photographs see Banesh Hoffmann, with Helen Dukas, *Albert Einstein: Creator and Rebel* (New York: Viking Press, 1972).

4. See Robert K. Merton, *The Sociology of Science: Theoretical and Empirical Investigations*, ed. Norman W. Storer (Chicago: Univ. of Chicago Press, 1973), and Joseph Ben-David, *The Scientist's Role in Society: A Comparative Study*, (Englewood Cliffs: Prentice-Hall, 1971), esp. chap. 7 on "German Scientific Hegemony. . . ."

5. For one aspect of "L'Académie Olympia de Berne," see the posthumously published correspondence of Einstein in *Lettres à Maurice Solovine* (Paris: Gauthier-Villars, 1956). For two famous accounts of Einstein's thought process, see Max Wertheimer, *Productive Thinking*, ed. Michael Wertheimer (New York: Harper, 1959), pp. 213–33, and Jacques Hademard, *An Essay on the Psychology of Invention in the Mathematical Field* (Princeton: Princeton Univ. Press, 1945), pp. 52, 142–43.

6. For sensitive guidance into these problems, see Gerald Holton, *Thematic Origins of Scientific Thought: Kepler to Einstein* (Cambridge: Harvard Univ. Press, 1973), pt. II, "On Relativity Theory," pp. 165–380.

7. For a collection of his early contributions to molecular physics, see *Albert Einstein, Investigations on the Theory of the Brownian Movement* (New York: Dover, 1956), ed. R. Furth, trans. A. D. Cowper; orig. ed., London, 1926, reprinting works from 1905 to 1908.

8. Gerald Holton, "Origins of the Special Theory of Relativity" in *Thematic Origins*, op. cit., p. 170.
9. Again, John T. Blackmore's biography, *Ernst Mach: His Work, Life, and Influence* (Berkeley: Univ. of California Press, 1972) is indispensable here.
10. Ernst Mach, *The Science of Mechanics: A Critical and Historical Account of Its Development*, trans. T. J. McCormack, 6th edn. from 9th German edn. (LaSalle, Ill.: Open Court, 1960), p. 279. "Mach's principle" appears on p. 316.
11. Ibid., pp. 347, 586.
12. H. A. Lorentz, *Collected Papers*, eds. P. Zeeman and A. D. Fokker, 9 vols. (The Hague: M. Nijhoff, 1935–39), vol. V, pp. 1–137: *Versuch einer Theorie der elektrischen und optischen Erscheinungen in bewegten Körpern* (Leiden, 1895). See also G. L. De Haas-Lorentz, ed., *H. A. Lorentz: Impressions of His Life and Work* (Amsterdam: North Holland, 1957). For discussions in depth of the Lorentz-Einstein relationship, see Elie Zahar, "Why Did Einstein's Programme Supersede Lorentz's?" *British Journal for the Philosophy of Science* 24 (1973): 95–123, and reactions by P. K. Feyerabend, A. I. Miller, K. F. Schaffner, *British Journal for the Philosophy of Science* 25 (1974): 25–78.
13. For Abraham, see the entry by Stanley Goldberg in C. C. Gillispie, *DSB*, vol. I, pp. 23–25. See also Goldberg, "The Lorentz Theory of Electrons and Einstein's Theory of Relativity," *American Journal of Physics* 37 (October 1969): 982–94. Cf. Carlo Giannoni, "Einstein and the Lorentz-Poincaré Theory of Relativity, in *Boston Studies in the Philosophy of Science* (Dordrecht: D. Reidel, 1970), vol. VIII, pp. 575–89.
14. R. Blondlot, *"N" Rays: A Collection of Papers Communicated to the Academy of Sciences*, trans. J. Garcin (London: Longmans, Green, 1905). Martin Gardiner, *Fads and Fallacies in the Name of Science* (New York: Dover, 1957), p. 345. Max Planck, *Where Is Science Going?* trans. James Murphy (New York: Norton, 1932), p. 78. Cf. Robert W. Wood, "The n-Rays," *Nature* 70 (29 September 1904): 530.
15. See Swenson, *The Ethereal Aether: A History of the Michelson-Morley Aether-Drift Experiments, 1880–1930* (Austin: Univ. of Texas Press, 1972), pp. 141–54.
16. Lord Rayleigh [John William Strutt], *Scientific Papers . . . by . . . Third Baron Rayleigh*, 6 vols. (Cambridge: Cambridge Univ. Press, 1899–1920). D. B. Brace, "On Double Refraction in Matter Moving through the Aether," *Philosophical Magazine*, 6th series, 7 (1904): 317–29.
17. F. T. Trouton and H. R. Noble, "The Mechanical Forces Acting on a Charged Electrical Condenser Moving through Space," *Philosophical Transactions* 202 (1903): 165–81; cf. E. H. Kennard, "The Trouton-Noble Experiment," *Electrodynamics of Moving Media: Bulletin of the National Research Council*, vol. 4, pt. 4 (Washington, 1922). Using a rotat-

ing Wheatstone Bridge several years later, Trouton and A. O. Rankine tried again to measure an aether-drift or a FitzGerald-Lorentz contraction, again without success; see their "On the Electrical Resistance of Moving Matter," *Proceedings of the Royal Society* (London) 80 (1908): 420–35.

18. See Tobias Dantzig, *Henri Poincaré: Critic of Crisis* (New York: Scribner's, 1954); Charles Scribner, Jr. "Henri Poincaré and the Principle of Relativity," *American Journal of Physics* 32 (September 1964): 672–78; Stanley Goldberg, "Henri Poincaré and Einstein's Theory of Relativity," *American Journal of Physics* 35 (October 1967): 934–44.

19. Henri Poincaré, *Science and Hypothesis*, trans. W. J. G. [sic] (New York: Dover, 1952), pp. 211–12; orig. edn. Paris, 1902.

20. Ibid., pp. 243–44. For a thorough analysis of the context here, see Arthur I. Miller, "A Study of Henry Poincaré's 'Sur la Dynamique de l'Electron,' " *Archive for History of Exact Sciences* 10 (1973): 207–328.

21. For Poincaré's address of September 1904 at the International Congress of the Arts and Sciences at St. Louis, see the translation by G. B. Halstead in *The Monist* 15 (January 1905): 1–24. See also the revision in Poincaré's *The Value of Science*, trans. G. B. Halstead (New York: Dover, 1958), pp. 91–111, orig. edn. Paris, 1905; also excerpted in L. P. Williams, ed., *Relativity Theory*, op. cit., pp. 39–49 (quotation here taken from p. 47).

22. A new translation with modernized notation is given by Herman M. Schwartz, "Poincaré's Rendiconti Paper on Relativity," in three parts, *American Journal of Physics* 39 (November 1971): 1287–1294; 40 (June 1972): 862–72; and 40 (September 1972): 1282–87, Cf. A. I. Miller's study cited above in note 20.

23. Planck's Nobel lecture, "The Genesis and Present State of Development of the Quantum Theory," for the Prize of 1918 but not delivered until 1920, is still one of the best sources for the "old theory": see *Nobel Lectures . . . Physics, 1901–1921* (Amsterdam: Elsevier, 1967), pp. 403–20. See also Max Planck, *Wissenschaftliche Selbstbiographie* (Leipzig: Barth, 1948), English trans. F. Gaynor, *Scientific Autobiography* (New York: Philosophical Library, 1949); and Thomas S. Kuhn, John L. Heilbron, Paul L. Forman, and Lini Allen, *Sources for History of Quantum Physics: An Inventory and Report* (Philadelphia: American Philosophical Society, 1967).

24. For helpful guides into quantum theory, see Gerald Holton, *Introduction to Concepts and Theories in Physical Science*, 2nd edn. rev. by Stephen G. Brush (Reading, Mass.: Addison-Wesley, 1973), esp. chaps. 26, 28, 29; and A. d'Abro, *The Rise of the New Physics*, 2 vols. (New York: Dover, 1951).

25. For Richardson, see article by Swenson in *DSB*, vol. XI, pp. 419–23.

26. F. Hasenöhrl, "Zur Theorie der Strahlung in bewegten Körpern," *An-*

nalen der Physik 15 (25 October 1904): 344–70; see also Philipp Lenard, *Great Men of Science: A History of Scientific Progress*, trans. H. S. Hatfield (London: Bell & Sons, 1954), orig. edn. Munich, 1933.

27. See J. C. Kapteyn's 1905 report "Discovery of the Two Star Streams" in Harlow Shapley, ed., *A Source Book in Astronomy, 1900–1950* (Cambridge: Harvard Univ. Press, 1960), pp. 105–11; also Kapteyn's "Recent Researches in the Structure of the Universe," a 1908 lecture, in Sir Bernard Lovell, ed., *The Royal Institution Library of Science . . . Astronomy*, vol. 2 (Amsterdam: Elsevier, 1970), pp. 78–96.

28. H. A. Lorentz, *Lectures on Theoretical Physics*, trans. L. Silberstein and A. P. H. Trivelli, 3 vols. (London: Macmillan, 1927–1931), vol. I, pp. 3–71, "Aether Theories and Aether Models" (1901–1902). See also Lorentz, *Collected Papers*, vol. V, pp. 172–97.

29. The standard English reference is the 1923 Methuen and Dover edition, *The Principle of Relativity: A Collection of Original Memoirs on the Special and General Theory of Relativity*, by H. A. Lorentz, A. Einstein, H. Minkowski, and H. Weyl, with notes by A. Sommerfeld, trans. W. Perrett and A. B. Jeffrey (hereafter cited as *The Principle of Relativity*), p. 38.

30. *The Principle of Relativity* (1905), pp. 48, 59.

31. Ibid. (1905), pp. 67–71.

32. Ibid. (1908), pp. 75–91.

33. For Minkowski's influence, see, e.g., Constance Reid, *Hilbert* (New York: Springer Verlag, 1970); Eugene Guth, "Contributions to the History of Einstein's Geometry as a Branch of Physics," in Moshe Carmeli, Stuart I. Fickler, Louis Witten, eds., *Relativity: Proceedings of the Relativity Conference . . . at Cincinnati* (New York: Plenum Press, 1970), pp. 161–207.

34. Einstein, "Fundamental Ideas and Methods of Relativity Theory, Presented in their Development," unpublished MS in Princeton Archives as presented by Gerald Holton in "On Trying to Understand Scientific Genius" in his *Thematic Origins of Scientific Thought*, p. 364. See also Cornelius Lanczos, *The Einstein Decade (1905–1915)* (New York: Academic Press, 1974).

35. See the Einstein bibliographies by Ernest Weil (1937; 1960) and Margaret C. Shields (1951) as listed in Bibliographical Note. A. Einstein, "Entwicklung unserer Anschauungen über das Wesen und die Konstitution der Strahlung," *Physikalische Zeitschrift* 10 (1909): 817–25. See also C. Lanczos, "Einstein's Path from Special to General Relativity," and N. L. Balazs, "The Acceptability of Physical Theories: Poincaré vs. Einstein," in L. O'Raifeartaigh, ed., *General Relativity: Papers in Honor of J. L. Synge* (Oxford: Clarendon Press, 1972), pp. 5–34. See again G. Holton's essay "On Trying to Understand Scientific Genius," pp. 356–62.

36. For more on Adler, see especially Lewis S. Feuer, *Einstein and the Gen-*

erations of Science, pp. 14–25. See also Carl Seelig, ed. *Albert Einstein und die Schweiz* (Zurich: Europa Verlag, 1952).

37. H. A. Lorentz, *The Theory of Electrons and Its Applications to the Phenomena of Light and Radiant Heat* (New York: Columbia Univ. Press, 1909), revised from lectures delivered at Columbia in March-April 1906, pp. 195–230; cf. 2nd rev. of 1915, reprinted by Dover in 1952.

38. H. A. Lorentz, "The Principle of Relativity for Uniform Translations," *Lectures on Theoretical Physics,* vol. III, p. 211, which dates from about 1910.

39. For C. E. Guye, see article by Swenson in C. C. Gillispie, *DSB,* vol. V, pp. 597–98. See also Russell McCormmach, editor's forewords, in *Historical Studies in the Physical Sciences,* annual vols. (Philadelphia: Univ. of Pennsylvania Press, 1969–   ), esp. vol. II, pp. ix–xx, and his article therein on "Einstein, Lorentz, and the Electron Theory."

40. Cf. G. Holton, "Poincaré and Relativity," in his *Thematic Origins,* pp. 185–95. Cf. Stanley Goldberg, "Poincaré's Silence and Einstein's Relativity: The Role of Theory and Experiment in Poincaré's Physics," *British Journal for the History of Science* 5 (1970): 73–84.

41. Stanley Goldberg, "In Defense of Ether: The British Response to Einstein's Special Theory of Relativity, 1905–1911," in R. McCormmach, ed., *Historical Studies,* vol. II, pp. 89–125. See also Dennis F. Miller, "The Early Influence of Einstein and the Special Theory of Relativity in America," unpublished M.A. thesis, Ohio State University, 1965.

Chapter V

The Einsteinian Synthesis
(c. 1910–20)

> Our question is "What are the generally covariant laws of Nature?"
>
> The special theory of relativity has led to the conclusion that inert mass is nothing more nor less than energy. . . . In the general theory of relativity we must introduce a corresponding energy-tensor of matter. . . .
> The general postulate of relativity is unable on principle to tell us anything about [the physical nature of matter]. It must remain to be seen, during the working out of the theory, whether electromagnetics and the doctrine of gravitation are able in collaboration to perform what the former by itself is unable to do.
>
> —ALBERT EINSTEIN, 1916

Our short perspective (only a century after his birth) that enables us to foreshorten slightly the full memory of Einstein's richness of personality, character, and thought also allows us to speak of his "synthesis" as a singular achievement comparable to Newton's or Maxwell's. Einstein's work, like Maxwell's, proved itself to be at the vanguard of professional physics by showing gradually a multiplex set of fruitful solutions to a wide variety of analytical problems. Thus, several syntheses—rather than merely one—lay between the special and general theories of relativity. The ongoing elucidation of relativity theory is and must remain unfinished because it deals with the cosmos as a whole, its origins and structure,

which we can "know" only in part through exploration and probabilities. Although man has just begun to probe nearby space in
person and with robot rockets, Einstein began more than half a
century ago to make the study of the universe as a whole scientifically and philosophically respectable once again.

Many of his forebears and contemporaries, especially experimental physicists, disdained such studies as "metaphysical," or worse,
as useless or meaningless diversions from whatever they considered the serious business of science. After 1919, however, when
Einstein's major predictions about *accelerated* relative motions began
to achieve observational and experimental corroboration, no one
could credibly maintain that science has no business or real ability
to speculate in a trustworthy manner about traditionally religious
questions concerning the cosmos.

With little doubt, this accomplishment was Einstein's greatest:
his syntheses paved the way to a revival of scientific cosmology
and cosmogony. He and such diverse contemporaries as Freud,
Cassirer, Whitehead, Russell, Shapley, Gandhi, and Schweitzer
breathed new life into man's faith in his rational capacities. He and
they also expressed fears for the future if man's irrational nature
should prevail. This double need for balanced pride and humility
in the face of the ambiguous nature of man and the contradictions
of human existence led to broadened definitions of science, nar-
rowed definitions of philosophy, and loosened definitions of religion.[1]

Few people could then fully appreciate these derivatives from
Einstein's thoughts and activities, largely because few had been exposed to experiences with natural objects on the exponential scales
necessary to deal with electrons and galaxies. Among those who
cared enough to master these problems of scale and manipulation,
however, the Einsteinian promise of new answers to cosmological
questions was eagerly welcomed. Beyond the European in-
telligentsia and even the intellectual world at large, Einstein's audience grew in reverence and gradually in understanding.

Many of his idealistic disciples came to see in him the promise
that some version of his "cosmic religious sense" might inspire a
growing minority to become a majority of enlightened mankind.
Some even dared hope that the Einsteinian synthesis might reunite

science and religion, for in his later years Einstein held fast to the possibility, however remote, of world government and human brotherhood. Einstein gained a charisma in many different areas as a "scientist, philosopher and contemporary conscience" whose pronouncements evoked enthusiastic followers and, inevitably also, some hostile critics. As recent biographers have remarked: "The essence of Einstein's profundity lay in his simplicity; and the essence of his science lay in his artistry—his phenomenal sense of beauty."[2]

All this became part of Einstein's mystique in the latter half of his life. Here we shall be mainly concerned to understand his maturity, how his professional status was ensured, what the professional competition for a general theory held in store, and where the implications of his conceptual critique led the profession. Then we shall look at three aspects of the Einsteinian syntheses in terms of their meaning for the working physicist. The proximate causes and cosmological effects of the advent of his general theory of relativity should help broaden our understanding of gravitation and of Einstein's lifelong attempts to unify the fields of physics for a scientific world-view. Finally, in an epilogue, we shall consider the evolution of the founder of the relativity revolution as a monistic rationalist philosopher, as a man of science *and* religion, as a man of "Spinoza's God."

Einstein's Maturity—To 1915

In his "Autobiographical Notes" written toward the end of his life, Einstein confessed, as honestly as he could recall, how the life of his mind had developed. In his continuous flights from momentary "wonders," he had tried to grasp some eternal verities about the huge world out yonder, "which stands before us like a great, eternal riddle, at least partially accessible to our inspection and thinking." Having relinquished religion for science and mathematics for physics, Einstein had found himself in the most fundamental of intellectual quests—seeking to match man's conceptual world to the totality of his perceived world of physical nature:

In a man of my type the turning-point of the development lies in the fact that gradually the major interest disengages itself to a far-reaching degree from the momentary and the merely personal and turns towards the striving for a mental grasp of things. . . .

I soon learned to scent out that which was able to lead to fundamentals and to turn aside from everything else, from the multitude of things which clutter up the mind and divert it from the essential.[3]

Since beginning his professional career, Einstein had been most fascinated by the electromagnetic researches of Maxwell and Hertz. He gradually came to see the electrodynamics of Faraday and Maxwell as creating a theory of fields of force that was more natural, comprehensive, and fruitful than the classical dynamics of Galileo and Newton. In fact, the force-fed motions of classical mechanics, which had necessitated mathematical distinctions between kinetics and kinematics as the concept of potential energy had to be incorporated into all branches of physics, seemed to him more and more artificial. Field theory explained action-at-a-distance more simply and completely than matter-and-motion theories had ever been able to.

Einstein came to see the pair Faraday-Maxwell as profoundly similar to the pair Galileo-Newton. He found most remarkable the intuitive grasp of the former of each pair and the synthetic power of the latter of each to formulate exactly the physical relationships between phenomena and conceptual understanding. Although he did not say so, the paired work of Lorentz and Einstein on theories of fields and electrons may now be seen to share this same inner similarity for the genesis of relativity. Equally significant, however, are the pair Planck-Einstein and their similar roles in applying discrete quanta to the continuum of radiation problems.

In September 1909 at the eighty-first assembly of German scientists meeting in Salzburg, Einstein had delivered his first invited paper before an audience of professional colleagues. Having chosen to speak on "The Development of Our Views on the Nature and Constitution of Radiation," Einstein strongly had expressed his belief that theoretical physics must move toward a "kind of fusion of the wave and emission theories" of light. This radical notion, anticipating the wave-particle duality that was to mature over the next two decades, had been a natural outgrowth of Einstein's work and attitudes since 1905. But Max Planck was not fully convinced by the cogent arguments of the young associate professor from Zurich. Planck arose first to comment conservatively that such a step surely ought to be avoided if possible. But other physicists in attendance, including Heinrich Rubens, Willy Wien, Arnold Sommerfeld, and Max Born, were clearly impressed by Einstein's arguments for fusing the wave and particle theories.[4]

The next year, in a joint effort with Ludwig Hopf, Einstein published two papers calculating momentum fluctuations that expanded and reinforced his earlier advocacy of wave-particle duality. Based on the 1905 papers on Brownian motion and electrodynamics, Einstein's views of the quantum structure of radiation were progressing toward duality. Despite the fact that wave theory was still needed to explain diffraction and interference phenomena, he felt the future would have to take light quanta as much more real than mere heuristic models.

From March 1911 until August 1912, Einstein served as a professor of physics at the German University of Prague. There he enjoyed an excellent library and the friendship of Georg Pick, who introduced him to the absolute differential calculus of C. G. Ricci and Tullio Levi-Civita. But life in Prague, with its intense antagonisms between Germans, Jews, and Czechs, was not idyllic, and Mileva was especially unhappy cooped up with her children. When the opportunity arose to return to Zurich and his alma mater, Einstein joyfully accepted the post of theoretical physics at the polytechnic institute where he had once failed to gain entrance.[5]

In the autumn of 1911 at Brussels, the Belgian industrialist Ernest Solvay convened a group of about twenty of the leading physical scientists from all over Europe to confer about certain critical problems besetting physics and chemistry. Walther Nernst made the arrangements, and Solvay played host to this distinguished gathering. It included H. A. Lorentz, Jean Perrin, Wilhelm Wien, Madame Curie, Henri Poincaré, Max Planck, Arnold Sommerfeld, Friedrich Hasenöhrl, James Jeans, Ernest Rutherford, Paul Langevin, Kamerlingh Onnes, and Albert Einstein. Although the theme of the conference focused on radiation and the quantum of action, Einstein delivered an invited paper on the state of studies of specific heats. It was well received, and this event marked Einstein's arrival into the elite of his profession.

What exactly was accomplished during the five days of discussions at the first Solvay conference is not clear, but the participants certainly must have profited from the interchange of ideas. Nernst, who seems to have patronized his benefactor Solvay for amateur theorizing, was spurred to pursue his notions for a heat theorem, which led into the so-called third law of thermodynamics about 1917. Rutherford, who was well known for his gruff empiricism and thorough experimentation, had already improved on J. J. Thomson's "plum-pudding" model of the atom, and his discovery of the nucleus would shortly inspire young Niels Bohr to improve on Rutherford's planetary model of the atom. If Lorentz, Poincaré, and Planck were the leading elder statesmen of this elite group, certain of the younger members such as Langevin, Jeans, Maurice de Broglie, and Einstein were surely en route to leadership. Einstein's preoccupation with studies of light and gravity, extending now beyond classical constraints and using new mathematical techniques, was becoming obvious to his peers. After this first Solvay conference, Einstein trusted his physical intuition to carry him from electrodynamics into gravitational theory.[6]

After moving from Prague back to Zurich, the Einsteins found themselves among many old friends. Most important, Marcel Grossmann, Einstein's mathematical comrade since student days, welcomed his return.

Grossmann had been studying non-Euclidean geometries and had thoroughly familiarized himself with the works of Riemann and Minkowski. He, too, was fascinated by the general problem of expanding the relativity principle into the realm of gravitational interactions. In correspondence Einstein had proposed some conundrums: Is gravity propagated finitely in time? Does the finite value for the velocity of light have an analogue with a finite velocity for gravity? If velocity is relative to so-called inertial systems, isn't acceleration also relative?

When we recall Maxwell's letter to Faraday in November 1857 asking the first (and last) of the great questions about gravity— "Does it require time?"—Einstein's quest for similar assurances seems in the mainstream of scientific development. But at the time of their collaboration in 1913, neither Einstein nor Grossmann had time for history. Rather, both were immersing themselves in the mathematical tradition of differential geometry stemming from Gauss and Riemann, Bolyai and Lobachevski. Grossmann resumed what Georg Pick had begun in introducing Einstein to the problems and prospects of the tensor calculus. This newly perfected mathematical technique, which had grown out of the theory of elasticity, allowed both the physical and the geometrical aspects of analysis to be preserved through the transformation from one coordinate system to another. Hence tensors became vital tools for the new theory of gravitation.

Having studied the culmination of Newtonian astronomy in Laplace's *Mécanique céleste* (published 1799–1825), Einstein was aware of Laplace's complaint that the scalar potential equation in spherical coordinates could not be integrated. This problem of the gravitational potential, as opposed to the kinetic energy of gravitational attractions, had given mathematicians much trouble since the eighteenth century. Poisson, d'Alembert, and George Green among others had worked on potential functions long before Kelvin, Stokes, Maxwell, Helmholtz, and Poincaré, again among others, wrestled with partial and ordinary differential equations, trying to find satisfactory ways to handle both the static and the dynamic aspects of field theory. Einstein's reverence for Lorentz and Minkowski and his reliance on Pick and Grossmann had led him even deeper into scalar, vector, and tensor analyses. Now, in 1913, he and Grossmann jointly published their first tensor treatment of gravity, but hardly had their collaboration been renewed before it again was ended.

At the apex of his intellectual powers, Einstein was recognized by his peers as a prize catch for any academic institution. But once happily resettled in Zurich, he might never have moved again had not Max Planck and Walther Nernst visited him in person in 1913 with a triple offer so flattering that he could hardly refuse to go to Berlin.

Planck and Nernst promised Einstein that he could remain a Swiss citizen if he came to the University of Berlin to fill the professional chair vacated at the death of J. H. van't Hoff in 1910. In addition, he was guaranteed a seat in the Royal Prussian Academy

of Sciences, and most enticing, Germany wanted Einstein to become the director of its new Kaiser Wilhelm Institute for Physical
Research. Despite the offer of three positions, his official duties
would be light, and the freedom and facilities to pursue his own
research would be virtually perfect. [7]

And so, in April 1914, Einstein at age thirty-five moved from
Zurich to Berlin. Still deeply preoccupied with his efforts to expand
his special relativity into a general theory that would account for
his insight into the gravity-inertia equivalence, Einstein quickly
adapted to the new environment and returned to work. But apparently, his wife and two young sons were not so happy, and after a
few months Albert and Mileva decided to separate. She and the
boys returned to Switzerland about the time Archduke Ferdinand
was assassinated at Sarajevo and the European balance of diplomatic power was shattered. Erwin Finlay-Freundlich, a young
Scottish-German astrophysicist, helped Einstein overcome this personal crisis and rebuild his professional momentum. Paul Ehrenfest, now of Leyden, was even more important as a confidential
friend, and Michele Besso, in Berne, looked after Einstein's family
commitments.

When the guns of August 1914 broke the peace of Europe and
the Pax Britannica, they also broke the quiet and contemplative life
of Germanic scholarship. Neither socialism nor science proved able
to practice the preachings of internationalism with the advent of
war. As diplomatic polarization gave way to nationalistic popularization during the first year of this first world war in a century or so,
Einstein cried out in anguish at those who deserted academic ideals
for patriotic duties. He took advantage of his Swiss citizenship to
denounce militarism and nationalism as hysterias. Ninety-three of
Germany's most celebrated intellectuals, including Max Planck,
signed a "Manifesto to the Civilized World," which attempted to
justify as pure self-defense the sweep of the Schlieffen Plan
through Belgium against France and the preventive aggression
against the "Russian hordes" on the eastern front. Einstein and
George Nicolai produced a counter "Manifesto to Europeans" in
the fall of 1914, calling for leading men of goodwill to transcend
their patriotism by declaring peace, unity, and support for a
League of Europeans. They could find only two others willing to

pledge and sign. As his political activities appeared more and more treasonous to his hosts, Einstein began to immerse himself ever deeper in his mathematics.[8]

Before the war began, Freundlich had set forth leading an eclipse expedition into the Crimea specifically to test Einstein's prediction that rays of starlight would bend in passing by the strong gravitational field of the Sun. That phenomenon was then observable only during a solar eclipse. The Russians arrested the German astronomers in August 1914 to prevent espionage, but within a month they were released in an exchange of prisoners.

Despite the disappointment of that first expedition to test the bent-starlight prediction, Einstein proceeded to work out the details of other testable assertions as necessary consequences from his theory. Already he had surmised that clocks should run more slowly in the neighborhood of ponderable masses; this implied a spectroscopic displacement of starlight toward the red end of the spectrum. Also a long-standing anomaly in the orbital behavior of the planet Mercury (its perihelion precessed too fast, by about 43 seconds of arc per century, for an explanation in accord with Newtonian mechanics) fitted nicely into the general theory's implications.

Gradually Einstein amplified these predictions, especially the Doppler-Fizeau principle applied to the "red shift" phenomena of distant stars. Even more gradual was his progress with mathematics for his new physics. The foundations of the general theory of relativity were being laid with the aid of the tensor calculus, according to the principle of the equivalence of gravity and inertia, and without the use of the concept of force. Einstein was trying to translate his vision of gravity as a property of space itself into a systematic set of field equations. But he was frustrated in his search for a logical way to deduce gravitational potentials.

In October 1915, over a year after the great European war had begun, Einstein working alone at a feverish pitch finally recognized the solution to his problem. This "eureka" experience was the peak moment of his life. To Arnold Sommerfeld he wrote an exultant letter on 28 November 1915, saying, "This last month I have lived through the most exciting and the most exacting period of my life and it would be true to say that it has also been the most fruitful."

In later retrospect, Einstein explained, "At last, in 1915, I recognized [my errors] as such and returned penitently to the Riemann curvature, which enabled me to define the relation to the empirical facts of astronomy." Thus did Einstein later describe how he cracked the nut he had been gnawing.[9]

His restricted relativity was valid only in space free of gravitational fields, that is, in microspace where gravity could be ignored, or in macrospace where astronomical events occur at such distances as again to allow one to ignore gravitational fields. Einstein concentrated on analyzing inertia, energy, and the gravitational field. By rejecting the circular reasoning involved in the notion of an "inertial system" as an explanation for the inertial behavior of matter, Einstein was able to incorporate three consequences that he had already guessed should follow from his theory. "The sense of the thing is too evident," he had told Besso early in 1914. Having brooded over these possible predictions since about 1908, once he recognized the significance of gravitational potentials and of the energy-tensor in expressing the symmetrical covariance of the gravitational field, Einstein was ready to write up the results of his labors and present them to the world of science. By 1916, that world, too, was torn asunder.

"The Foundations of General Relativity Theory"

In correspondence with friends and colleagues, in visits to Switzerland and Holland, and in many preliminary papers, Einstein had paved the way for his general theory of relativity. *Annalen der Physik* and the Prussian Academy claimed first rights for publication. In the midst of a war growing more bitter day by day, Einstein published (1916) a sixty-four–page offprint through the firm of J. A. Barth in Leipzig and this became the primary vehicle of its international communication. Entitled in German *"Die Grundlage der Allgemeinen Relativitätstheorie,"* this pamphlet reached out beyond Germany to the scientific societies of other warring states where it was received with considerable anticipation by those few who could appreciate its message. Characteristically, the theme of Einstein's pamphlet-paper on "The Foundations of General Relativity Theory" was embedded in his quest for unity, simplicity, generality, for some stability in the midst of flux: "Our question is 'What are the generally covariant laws of nature?' "[10]

Einstein's *Grundlage* paper was his *magnum opus*, the great work that both he and his disciples (and enemies) considered the capstone of his career. Short as it was, it represented more than a decade of concentrated personal effort to match mathematics and physical nature in a coherent, corresponding model of the space-time continuum. Thanks to his previous critiques of the basic concepts of space, time, mass, energy, force, and inertia, Einstein now extended and systematized the field concept to encompass the whole universe in a new theory of gravitation. The penultimate section of this paper included Newton's theory of universal gravitation as a first approximation (wherever the motions of matter generating the gravitational field are slow enough to ignore, in comparison with the velocity of light), within the more comprehensive system of equations offered by Einstein.

Einstein organized "The Foundations" by presenting it in five major parts divided into twenty-two sections. Beginning with some fundamental considerations on the postulate of relativity as embedded in the mechanics of Galileo and Newton, Einstein observed how his special theory of relativity grew out of the Lorentzian tradition when he had postulated the constancy of the velocity of light *in vacuo*. Now the need for an extension of the postulate of relativity became clear after considering Ernst Mach's objections to Newtonian notions of absolute space, time, and motion. To get closer to observable facts of experience, as opposed to factitious or artificial conventions regarding the causes and effects of relative motion, Mach and now Einstein argued for a purification of scientific judgments by recognizing the relativity of inertial and gravitational masses. *"The laws of physics must be of such a nature that they apply to systems of reference in any kind of motion."*[11] In effect then, Einstein's thought had evolved from the special theory of relativity, with its restrictions to uniform rectilinear motion and the constancy of the speed of light, toward his general theory of relativity, with its accommodations for uniformly accelerated motions (or translations, rotations, nonlinear directions, and so forth). By focusing on the remarkable basic fact of experience that *all bodies* regardless of composition, density, weight, form, or shape *fall freely* in a gravitational field devoid of air or other resistances *with the same acceleration,* Einstein was able to see how the principle of the constancy of the speed of light must be modified for encounters with strong gravitational fields and how these curvilinear paths of rays of light lead into a better theory of gravitation.

Taking the space-time continuum seriously, Einstein had to give up hope of finding the center of the universe or of measuring with rigid rods or standard clocks any definitive aspect of a "stationary" body. What is "really" at rest or in motion ultimately depends upon what coordinates or reference frame you choose to define as your point of origin:

So there is nothing for it but to regard all imaginable systems of co-ordinates, on principle, as equally suitable for the description of nature. This comes to requiring that:—

The general laws of nature are to be expressed by equations which hold good for all systems of coordinates, that is, are co-variant with respect to any substitutions whatever ([that is, they must be] *generally co-variant*). [12]

In introducing the mathematics of his new-found metric system (the "four co-ordinates to measurement in space and time"), Einstein confessed that his main object was to develop his theory in the most persuasive and convincing form rather than in the simplest and most logical manner. Whatever he may have meant by this declaration of purpose (p. 118), it is clear that he assumed his audience would be primarily theoretical physicists rather than mathematicians or astronomers.

The second major part of "The Foundations" is entitled "Mathematical Aids to the Formulation of Generally Covariant Equations." This part is twice the length of any other part and yet has an anecdotal rather than a systematic character. Introducing the tensor calculus, the equation of the geodetic line, and the Riemann-Christoffel tensor, among other "aids," Einstein may have felt constrained not to trespass on the concurrent work of his senior colleague in mathematics, David Hilbert in Göttingen. At the end of his third part, the core, "Theory of the Gravitational Field," Einstein gives Hilbert a footnote of credit; some think he deserves more.

Because this third part of "The Foundations" quickly became and remains the most influential, a few more extended quotations from it may convey the flavor if not the meat of its message:

> We make a distinction hereafter between "gravitational field" and "matter" in this way, that we denote everything but the gravitational field as "matter." Our use of the word therefore includes not only matter in the ordinary sense, but the electromagnetic field as well.[13]

And shortly thereafter Einstein expresses in words what his mathematics had finally made clear to him:

> The special theory of relativity has led to the conclusion that inert mass is nothing more nor less than energy, which finds its complete mathematical expression in a symmetrical tensor of second rank, the energy-tensor. Thus in the general theory of relativity we must introduce a corresponding energy-tensor of matter . . . which, like the energy-components . . . of the gravitational field, will have a mixed character, but will pertain to a symmetrical covariant tensor. . . .
>
> This introduction of the energy-tensor of matter is not justified by the relativity postulate alone. . . . The energy of the gravitational field shall act gravitatively in the same way as any other kind of energy. . . . The equations of conservation of momentum and energy . . . hold good for the components of the total energy.[14]

At the start of Einstein's fourth part, on "Material Phenomena," he admits that no new principle of impotency, comparable to that against absolute motion in the special theory, has been derived so far from the general theory. But he boasts that the ubiquitous influence of gravity on all processes may now be accounted for without any new hypotheses:

Hence it comes about that it is not necessary to introduce definite assumptions as to the physical nature of matter. . . . In particular it may remain an open question whether the theory of the electromagnetic field in conjunction with that of the gravitational field furnishes a sufficient basis for the theory of matter or not. The general postulate of relativity is unable on principle to tell us anything about this. It must remain to be seen, during the working out of the theory, whether electromagnetics and the doctrine of gravitation are able in collaboration to perform what the former by itself is unable to do.[15]

Having handled Euler's hydrodynamic and Maxwell's electrodynamic equations in terms of his new theory, Einstein turns in his fifth and last untitled part of "The Foundations" to Newton's theory as a first approximation for universal gravitation and to Euclid's geometry likewise:

Thus Euclidean geometry does not hold even to a first approximation in the gravitational field . . . although, to be sure . . . the deviations to be expected are much too slight to be noticeable in measurements of the earth's surface.[16]

In conclusion, Einstein reiterates three operationally testable predictions that for several years now had needed this more complete mathematical treatment to assert with confidence:

[I.] Thus the clock goes more slowly if set up in the neighborhood of ponderable masses. From this it follows that the spectral lines of light reaching us from the surface of large stars must appear displaced toward the red end of the spectrum. . . .
[II.] The course of [star] light-rays must be bent. . . . A ray of light going past the sun undergoes a deflection of 1.7"; and a ray going past the planet Jupiter a deflection of about .02". . . .
[III.] We find a deviation . . . from the Kepler-Newton laws of planetary motion. The orbital ellipse of a planet undergoes a slow rotation, in the direction of motion.[17]

By concentrating on discursive words rather than the deductive mathematical content, these excerpts from Einstein's major treatise give only hints of the power of thought contained therein. Because that predictive power was soon demonstrated through empirical proof of his predictions, Einstein was catapulted to the pinnacle of honor as a scientist. We shall now return to Einstein's three major conceptual syntheses—mass-energy, space-time, and gravity-inertia—to retrace their evolution between 1905 and 1915 in more detail.

The Mass-Energy Identity

As we saw in previous chapters, the many trends—physical, philosophical, psychic—that came together in Albert Einstein's life and work were anticipated in almost every particular long before 1905. The specific strands of the network of Einstein's scientific attitudes in 1905—toward molecular reality, Planck's quanta of action, and the principle of relativity—were soon embroidered elaborately into a more general tapestry of physical science. Einstein's own question "Does the inertia of a body depend upon its energy-content?" and his affirmative answer already in 1905—"The mass of a body is a measure of its energy-content"—showed immediate promise of some profound conceptual consolidations.[18]

Because Einstein had excluded from his analysis of the electrodynamics of moving bodies all considerations of nonuniform motion, he had also excluded gravitational phenomena. This meant that he was dealing here only with inertial mass and motions in straight lines, not with accelerations or rotations. Hence, classical ambiguities in the concepts of mass and inertia were brought to the forefront of attention as Einstein and others wrestled with the conservation laws of momentum, energy, mass, and inertia. The relativistic notion of inertial mass was clarified by Hermann Minkowski in 1908 and by Gilbert N. Lewis and Richard C. Tolman in 1909.[19]

Ever since 1881, when J. J. Thomson had first envisioned the likelihood of relating inertia to Maxwell's synthesis of electromagnetism, theoreticians had toyed with various combinations for relating mass to energy. J. H. Poynting, W. K. Clifford, Joseph Larmor, H. A. Lorentz, Wilhelm Wien, Max Abraham, Georg Helm, Gustav Mie, and Henri Poincaré were among those who had anticipated Einstein in these respects. The etherialization of matter and the reification of the field concept had proceeded so far by 1900 that the electromagnetic world-view seemed about to engulf theoretical physics. Max Jammer has expressed this intellectual change in physics and philosophy as follows: "Matter does not do what it does because it is what it is, but it is what it is because it does what it does."[20]

Curiously, Einstein's own first derivation of the famous formula $E = mc^2$ was incorrect in the sense of begging the question of what was to be proved. Growing out of Einstein's subliminal obsession with the operational meaning of the constancy of the velocity of light, the mass-energy equivalence

$$m = \frac{E}{c^2}$$

had been assumed in interior calculations as

$$\frac{E}{mc^2} = 1$$

and thus the equivalences

$$E = mc^2$$
$$\frac{E}{c^2} = m$$

and

$$\sqrt{E/m} = c$$

were embedded in the premises, therefore predetermined in the conclusion. Though right for the wrong reasons at first, Einstein caught his mistakes and redressed his deductions in further publications in 1906 and 1907.[21]

Meanwhile, such monistic and positivistic philosophers as Hermann Minkowski, Gustave Le Bon, Julius von Olivier, and Josef Petzoldt seized on these theoretical results identifying mass and energy as different names for the same thing. As different aspects of an underlying physical reality, could these concepts be generalized by asserting matter or mass to be merely frozen energy; and energy, whether kinetic or potential, to be merely mass or matter in process of evaporating, condensing, or awaiting the opportunity to do so?[22]

Over the next three decades, experimental physicists slowly but surely began to build up the evidence through atomic, nuclear, and particle physics for the evolution of the mass-energy equivalence into the mass-energy identity. Meanwhile, Einstein, Planck, Paul Langevin, and others built up the empirically based theoretical edifice wherein the laws of the conservation of mass from chemistry coalesced with the conservation of energy from physics.

Hence, by 1913 or so, the relativistic notion of the mass-energy identity was embodied in the new synthesis of the law of conservation of mass-energy, or "massergy" as Jammer suggests. Steadily, between 1905 and 1915, Einstein's confidence, despite spurious experimental results to the contrary, was reinforced by his vision of the breadth and depth of the cross-fertilization of these concepts. Others perhaps would have to wait until 1933, when P. M. S. Blackett and G. P. S. Occhialini learned to create and annihilate pairs of elementary particles in the laboratory. No one could deny the mass-energy identity after the creation and destruction of electron-positron pairs.[23]

But by then Einstein's fame for the general theory outshone his work in the special theory of relativity. As he fled Europe and the Nazi menace for America and the tranquillity of Princeton, New Jersey, Einstein found himself removed from the main community of natural philosophers who had taken such matters seriously so far. But he also transplanted that tradition to Princeton, attracted a new community to its Institute for Advanced Study, and sowed seeds of interest that flourished there in several forms, nourishing, for example, John Archibald Wheeler's geometrodynamics.[24]

The Space-Time Nexus

Although Einstein's denial of absolute space and time as inherited from the Newtonian tradition was implicit in his critique of the concept of simultaneity, his explicit theory as set forth in the several papers of 1905 and 1906 was preoccupied with time rather than space, with physical rather than geometrical problems. Having raised the constancy of the velocity of light from an empirical generalization to an unfalsifiable postulate, Einstein had committed himself to working out the details for a physically meaningful, operationally testable theory of electromagnetism and electronics. He adapted Lorentz's transformation equations without the deformable electron, together with Poincaré's emphasis on the principle of relativity and the unique character of the velocity of light, to elaborate his intuitive notions about the basic concepts of physics. But only after he had worked out to his own satisfaction the relationships between mass, inertia, and energy did he turn his attention toward the concepts of space and gravity.

In 1907, at Johannes Stark's request, Einstein wrote a long article for the German *Yearbook of Radioactivity and Electronics* entitled "The Principle of Relativity and Its Consequences," which dealt with optics, electrodynamics, experimental electronics, thermodynamics, the conservation laws for mass and energy, and the principle of relativity in relation to gravitation. By then, Einstein had become thoroughly convinced that physics was moving toward the acceptance of fields and particles as more real than forces and material objects. As I mentioned in the last chapter, Einstein called his "eureka experience" in 1907 that gravitational fields, like electromagnetic fields, have only relative existences (his discovery that gravity and inertia are ultimately indistinguishable) the "happiest thought" of his life.[25]

This gravity-inertia linkage quickly became known as his principle of equivalence, and it grew directly into the main idea of Einstein's theory of gravitation. But en route, even in 1907, Einstein had to restrict his postulate of the invariance of the velocity of light in order to generalize the postulate of relativity. Because the process of generalizing the principle of relativity, which had so far been restricted to coordinate systems in *uniform* rectilinear motion, suggested that it might hold equally well for reference frames in uniformly *accelerated* translational motion, Einstein found himself calculating that light beams passing through gravitational fields must be bent. Rigid rods and standard clocks must also be affected by strong gravitational potentials. Considering the variability of standard measuring devices for space and time intervals led inevitably into reconsidering what are meant by invariance, covariance, contravariance, by metric, manifold, and "rest mass." In short, during the four years between 1907 and 1911, Einstein, while wrestling mostly with quantum phenomena and the perplexities of time, evolved into being profoundly concerned with the enigma of gravitation, space at large, and the relations of geometry to physics.[26]

Without doubt, the mathematician Hermann Minkowski was most influential in bringing Einstein into prominence and encouraging him to rethink the implications of the principle of relativity. During 1908, the year before he died prematurely, Minkowski launched a campaign to merge the concepts of space and time. As a representative of Göttingen, the mathematical Mecca of Germany where Felix Klein, David Hilbert, and Emmy Noether reinforced each other, Minkowski spoke strongly for the creation of a new mathematical physics. He argued that "space by itself and time by itself, are doomed to fade away into mere shadows" and that only a new marriage of the two could be expected to "preserve an independent reality." In his famous address at the Cologne convention on 21 September 1908, Minkowski challenged all physicists and mathematicians to undertake the "unavoidable" revision of all physics on the basis of four dimensions. Crediting Michelson's experiment, Lorentz's electronic theory, and Einstein's critique of time and simultaneity, Minkowski tried to convert the relativity postulate into a "world-postulate." Instead of a singular concept of space, he showed how an infinite number of *spaces* (plural) could be derived from the concept of a four-dimensional space-time continuum. "Three-dimensional geometry becomes a chapter in four-dimensional physics" (p. 80). Professing thus to have simplified Newtonian mechanics and modern electrodynamics by having unified space-and-time, Minkowski concluded that both Newtonian gravitation and Lorentzian electromagnetism could eventually be unified in a unitary field theory for the whole of this unique ubiquitous universe.[27]

By 1911, when in Prague Einstein began to expose himself seriously to non-Euclidean geometries, he returned to the question of the influence of gravitation on the propagation of light. His paper on the subject published that year drew caustic criticism from Gunnar Nordström, a Finn from Helsinki who was then in Göttingen. For different reasons both Abraham and Nordström wanted to preserve the constancy of the velocity of light, which Einstein was now modifying to save the principle of relativity. Forced to defend himself, Einstein was also forced to clarify and extend his principle of equivalence and to probe more deeply the space-time nexus, that lay waiting in differential geometry and the philosophy of mathematics.

Einstein went back to Leibniz, Huygens, Kant, and Gauss to try to understand why Euclidean geometry and Newtonian physics were proving inadequate in the realms of nuclei and nebulae. Visiting Mach in Vienna and conferring with Poincaré at the first Solvay conference in late October 1911, Einstein gained reassurance that neither of these grand old men (representing neo-positivism and conventionalism, respectively) was necessarily closer to the nub of the problem of analyzing the space-time nexus than he was himself. Consequently, on his return to Zurich and to collaboration with his friend Marcel Grossmann in September 1912, Einstein immersed himself in the effort to understand how best to reconcile abstract geometries with experiential physics.[28]

During 1912–13, Einstein and Grossmann were trying to find the appropriate forms of the absolute differential calculus to apply to general relativity and the new theory of gravitation. They must have been at least dimly aware that not far away, at the university in Göttingen, Gustav Mie, Gunnar Nordström, Emmy Noether, and David Hilbert himself, one of the greatest active mathematicians, were all working on similar problems.

Likewise in Paris, Poincaré had turned his attention toward topology and was, in what was to be the last year of his life, trying to prove through an analysis of the concept of dimensions that physical space is most conveniently and coherently to be described as Euclidean, three-dimensional. Even Luitzen E. J. Brouwer (1881–1967) at the University of Berlin, the chief exponent of the intuitionist school of thought (as opposed to Hilbert's formalist school and Bertrand Russell and Alfred North Whitehead's logistic school), was known at least to Grossmann as a competitor in the race for a new gravitational theory. Brouwer, who believed, as had William Rowan Hamilton, that algebra is the science of time as geometry is that of space, was during this same period giving birth to his topological invariance theorems.[29]

And so, aside from the friendly rivalry between mathematician Grossmann and physicist Einstein as they cooperated on the quest for a new gravitational field theory, others in Europe were also hot on the trail of a new set of gravitational field equations. The philosophical and methodological attitudes of the contestants differed so much, however, that quite different aspects were to be emphasized by each school of metamathematics and metaphysics.[30]

If Einstein's initial insight in special relativity was the realization that it is impossible to synchronize distant clocks because light does not travel instantaneously, it may be said that his central quasi-mathematical insight in arriving at his general theory was the notion that the curvature of the space-time continuum could be trusted to produce geodesics that would satisfy the conservation laws of momentum and energy. The general theory gave no new principle of impotency, but it did link up a complex set of theoretical constructs in such a consistent and noncontradictory system that its main goal, the field theory of gravity, promised to become also the basis for a field theory of everything else. Space-time was the mediating synthesis, both chronologically and psychologically, between the restricted and general theories of relativity.

The Gravity-Inertia Equivalence

Einstein and Grossmann published a report of their progress as an "Outline of General Relativity and of a Theory of Gravitation," then listened for reactions. Mie, Abraham, and Nordström were quick to criticize in public; Lorentz, Langevin, and Sommerfeld were rather more encouraging in private. Meanwhile, experiments were making news, too. Ernest Rutherford's laboratory in Manchester and James Franck's in Berlin were forging evidence of the existence of atomic nuclei and protons to complicate the problem of defining "matter." Einstein, watching both kinds of news, felt he was finally becoming sophisticated enough in physics and mathematics to stand and fight his own battles.

The secret of his self-confidence was his memory of the thrill that his central insight into this principle of equivalence had given him. Because all bodies regardless of their material composition fall freely in a vacuum and with the same acceleration (as proved again most recently by the 1908 experiments of Baron Roland von Eötvös), surely large, accelerated systems cannot be distinguished ultimately from gravitational systems; therefore, the laws of inertia and of gravity can be treated as equivalent. If inert mass is latent energy and if the time dimension may be treated formally on equal terms with the three space dimensions, surely the generalized theory of relativity should incorporate these earlier syntheses. This could lead to a better theory of gravitation, even if that meant the abandonment of Euclidean geometry and the modification of the Newtonian laws of motion.[31]

In late 1913, Einstein in Zurich was busily at work on many papers in theoretical physics when he was invited to move to Berlin. He was not willing to relinquish his Swiss citizenship, but Planck and Nernst together with their colleagues Heinrich Rubens and Emil Warburg were able to negotiate a settlement. And so, as we have seen, the Einsteins moved to Berlin in April 1914. As the carnage of August 1914 spread over Europe, the Einstein family was broken asunder. Mileva and the boys returned to the beauty and neutrality of Zurich, leaving Albert to his studies and new-found colleagues. The Berliners were most congenial scientifically but hardly civilized, in Einstein's view, because their hostility and nationalism outweighed their humanism and loyalty to Europe.

Nevertheless, during 1915, Einstein concentrated as never before on the analytic problems lying at the basis of the grand new synthesis he was trying to create. Striving for logical perfection and secure foundations in a physical (as opposed to a mathematical) sense, Einstein sought through his calculations to reconcile the problem of the gravitational potential energy with certain field equations expressing both general covariance and the equivalence principle.[32]

Meanwhile David Hilbert and his circle in Göttingen were approaching the "axiomatization of physics," the fifth of Hilbert's famous set of twenty-three problems for the future of mathematics, which he had set forth in Paris in 1900. Hilbert had deliberately chosen to base his formalistic approach on Gustav Mie's field theory of matter and on Einstein's special theory of relativity. While most of the members of the scientific professions were being mobilized for war work, Hilbert and Einstein remained free to pursue their muses. Hilbert used the variational calculus and the theory of invariants, powerful tools of which he was the world's past master. Einstein used the tensor calculus and a theory of covariant and contravariant four-vectors. Hilbert felt that Einstein's approach was too messy and not closely enough meshed with the electromagnetic theory of matter to lead into a unified field theory. Einstein felt that Hilbert's axiomatic methods were too elegant and not yet justified at their level of generalization, while there was still so much ferment in experimental physics regarding the nature of "matter" or electromagnetic fields and particles.

Apparently both Einstein and Hilbert independently experienced their "eureka" insights in October 1915, and during the next few weeks they separately polished their gravitational field equations for public presentations. Hilbert spoke to his academy in Göttingen on 20 November 1915; Einstein to his academy in Berlin on 25 November 1915.[33]

Publication of their results early the next year might have led to priority fights except for the principal investigators' mutual esteem and admiration. Felix Klein and the brilliant young Wolfgang Pauli, Jr., were sympathetic to Hilbert, but Planck, Lorentz, Nernst, et al. were fairly partisan toward Einstein. Thanks largely to H. A. Lorentz's adjudication during 1916, the equivalence of the Einstein-Hilbert field equations was accepted by the profession. As a leading scholar of the episode has affirmed, "Altogether Lorentz had produced a complete proof of the equivalence of Einstein's inductive and Hilbert's deductive methods, treating all of the delicate points clearly and in detail."[34]

David Hilbert *(courtesy AIP Niels Bohr Library—Landé Collection)*

Niels Bohr *(courtesy AIP Niels Bohr Library—Fermi Collection)*

Astrophysicists at Mount Wilson—*left to right:* Humason, Hubble, St. John, Michelson, Einstein, Campbell, and Adams—under a portrait of Hale (*courtesy Hale Observatories Library—Michelson Museum*)

Other genteel disputes over the best interpretations to be placed on the foundations and the superstructure of the completed relativity theory arose quite quickly also, despite the limitations imposed by the war of attack turning into a war of attrition.

Einstein's latest synthesis not only denied gravitation *as a force* but also as an *instantaneous* propagation of action-at-a-distance. Because force-fed motion is dynamical by definition, and an "instant" is also an absolute, there was room for much debate. Einstein's abstract kinematics, incorporating the four-dimensional space-time continuum, also encompassed Newton's laws of dynamics. The gravity-inertia equivalence now meant abandoning "straight lines" in favor of "geodesics" as the shortest distance between two points. Although Mach and Hertz had argued years earlier that there is no way ultimately to distinguish between the influence of "hidden masses" at large in the universe and the centrifugal and centripetal "forces" of gravity, the concepts of mass, energy, space, time, inertia, and gravity were only now being fused into an entirely different world-view.[35]

So provocative was Einstein's critique of these interlocking concepts from classical mechanics that his basic presuppositions had to be scrutinized over and over again. Hilbert had supplied the contemporaneous bases for a cultivated uproar over the mysterious "fourth dimension" and "geometrization" of physics. Einstein consequently felt the need to clarify for all audiences how mathematical physics (à la Hilbert et al.) differed from his own theoretical physics. In a 1917 article-pamphlet, his *"Gemeinverständlich,"* Einstein set forth discursively to explain the basic postulates of general relativity as simply three main principles recombined with a new emphasis: first, the principle of relativity itself from 1905; second, the principle of equivalence for inertia-gravity from 1907; and third, the so-called Mach's principle (local phenomena of inertial systems are determined by all the masses of the universe and their distribution) from around 1883. These three principles had often been confused, and so the general theory of relativity had been designed especially to distinguish between the first and last principles. The laws of nature purport to state coincidences in space-time and therefore must be covariant. Machians had argued to destroy the monstrosity of "absolute motion" that the gravitational behavior of

a body, whether being "attracted" to some other "center of gravity" ("centripetal force") or being "repelled" from some other
"center of gravity" ("centrifugal force"), might with equal justice be
determined by the interrelatedness of the whole universe of matter.
Who could say whether the Earth system, solar system, galactic
system, and so forth followed in their super-celestial mechanics the
notions laid down in Newton's synthesis? Centripetal and centrifugal forces *as forces* may well be illusions. Ultimately the mass-
energy relations of the cosmos become manifest to us in local rep-
resentations of various gravitational fields.[36]

By separating these philosophical assumptions in his mathematical treatment, Einstein obtained two fresh sets of laws, one of
structure and one of motion. He felt he had stayed true to physics
without compromise to the "geometrizational" tendency or exces-
sive formalism, which he deplored. Most practicing physicists,
however, found this attitude hard to understand at first. Each one
who tried to understand soon seemed to have to write a book on
the subject in order to convince himself at least that he understood.
By now Einstein knew and used more mathematics than most of
his fellow physicists; yet he always insisted that he was primarily
one of them and not a mathematician.

Two events in 1916 piqued Einstein's social conscience and
psychological self-consciousness. In the midst of the dark and bitter
warfare, an old friend and fellow student of physics became a political assassin: Friedrich Adler shot and killed the Austrian prime
minister, Count Karl von Stürgkh, on 24 October 1916. A
privileged son of a distinguished family, Adler had turned Machian
in philosophy and Marxist in politics, but his singular act of protest
led to much traumatic grief and to no apparent shortening of the
war. Einstein had owed his first chair in Zurich in 1908 in part to
Adler's charity; now he felt he had to try to help save the life of a
murderer.[37]

The second revelation was less painful but equally probing. Max
Wertheimer, a psychologist of the Gestalt school, started a series of
interviews in depth, trying to get Einstein to dredge up the
mainsprings of his creative-thinking processes. The experiment
may not have been as productive as Wertheimer hoped, but it provided Einstein with valuable insights into himself.[38]

In 1917, during one of Einstein's rare visits to Switzerland to see their sons, Hans (then thirteen) and Eduard (then seven), Albert and Mileva decided to divorce as soon as possible after the war. Meanwhile, Einstein had learned to depend on his widowed cousin Elsa during the cold and hungry last years of the war. Chronic stomach trouble began to plague him from then onward. And yet, thanks to certain collaborators such as Otto Stern, A. D. Fokker, and W. J. de Haas, Einstein remained stable and focused in his work. Indeed, he produced another synthesis in 1917, which linked radioactive disintegration with Planck's theory and thus laid another major foundation stone for the new physics.[39]

Gravitation and Cosmology

Einstein remained prolific throughout the war, averaging almost eleven papers per year and paying virtually equal attention to macrophysics and microphysics. But quantum, atomic, and molecular topics were no longer his forte, as general relativity, which now entirely incorporated gravitation, beckoned for further extensions. Having made his peace with Lorentz and Hilbert by respectfully insisting on his own point of view, Einstein next published a paper (1917) that moved beyond the "planetary problem" (or the solar-system framework) and into "larger portions of the physical universe."

Entitled "Cosmological Considerations on the General Theory of Relativity," this work examines the Newtonian system of the world to find certain paradoxes regarding the mass-energy distribution that is both finite and infinite, from different points of view toward the limits of our vision. In examining the boundary conditions associated with the general theory, Einstein confessed his recent change of mind regarding the gravitational field equations introduced in 1915–16. Then he had assumed without looking far afield that "there can be no inertia *relatively to 'space'* but only an inertia of masses *relatively to one another.*" Now, however, in reconsiderations (thanks to Jacob Grommer and Willem de Sitter), after having tried to allow the gravitational potential to degenerate to zero in the far reaches beyond the so-called celestial sphere of fixed stars,

Einstein wished to introduce a new symbol ($-\Lambda$) for a cosmological constant. This additional term for the field equations would ensure equilibrium for the flexible yet quasi-static distribution of matter as observed in the relatively small proper motions of stars. Here was presented the notion of a finite but unbounded spherical continuum with positive curvature.[40]

Einstein wanted to ensure that his model of the universe on the largest conceivable scale should be spatially finite with a virtually uniform distribution of matter. But such speculations brought forth immediate criticism and counterspeculations from Hermann Weyl, Gustav Mie, Hilbert, Sitter, and others. In fact, Weyl's famous book on *Space-Time-Matter* was first published within the year as part of another alternative model for a unified field theory. But this geometrization of physics, though elegant, was not empirical enough to be satisfying to Einstein's physical intuition. At least to him, mathematicians tended to generalize too far too fast, without enough attention to astrophysics.[41]

Einstein returned to attack the geometricians invading physics in his 1919 paper titled with the question "Do Gravitational Fields Play an Essential Part in the Structure of the Elementary Particles of Matter?" The first equation herein came to symbolize the form for Einstein's field equations thereafter:

$$G_{\mu\nu} - \tfrac{1}{2} g_{\mu\nu}G = -\kappa T_{\mu\nu}$$

"where $G_{\mu\nu}$ denotes the contracted Riemann tensor of curvature, G the scalar of curvature formed by repeated contraction, and $T_{\mu\nu}$ the energy-tensor of 'matter.' " Compressed into this statement of equality was the whole corpus of Einstein's relativity theories worked out over fourteen years. And although he was later to feel that the left-hand side of the equation was "made of fine marble," while the right-hand side was "built of low-grade wood," in this 1919 paper Einstein manipulated the left-hand side, and particularly the first symbol of the second term, in order to prove "the possibility of a theoretical construction of matter out of gravitational field and electromagnetic field alone, without the introduction of hypothetical supplementary terms."[42] Einstein was elated to have avoided the need for that cosmological constant ($- \Lambda$), but he was sorry to answer that elementary quanta had not yet yielded to the given field equations.

With the Armistice of November 1918 came the "peace without victory" that quickly turned into the *Diktat* of Versailles inside tumultuous Germany. In January 1919, Einstein went to Switzerland to dissolve his first marriage officially. On June 2, in Berlin, he married Elsa Lowenthal and acquired two stepdaughters, Ilse (twenty-two) and Margot (twenty). If these changes in his domestic life disturbed his work at first, they eventually brought him more tranquillity. He remained aloof from possessive and familial emotions, and this coldness he admitted as a fault.[43]

Einstein was forty years of age in 1919 when, amid the turmoil and hunger in defeated central Europe, a British naval expedition carried two groups of British astronomers to the mid-Atlantic to make another attempt to gather photographic evidence during a solar eclipse and thereby test Einstein's prediction of the bending of starlight as it passes through a strong gravitational field. Sir Frank Dyson had been planning such a test since March 1917. Arthur S. Eddington, a Quaker astronomer strongly in sympathy with Einstein's scientific as well as sociopolitical attitudes, was in charge of the eclipse observations to be made on the island of Principe in the West African Gulf of Guinea. The other group went to Sobral, a town in northern Brazil between Belém and Recife, preparing their instruments for the fateful 29 May. The weather at both sites that day was less than ideal, but Eddington's group took sixteen photographs on faith during the five minutes of total eclipse. When later seven plates from Sobral were compared with the best from Principe, Eddington became jubilant to find that *"light* has *weight"* and "light-rays [passing] near the sun *do not go straight."*[44]

Official announcements of these results were delayed until final checking could be completed in early September. Lorentz telegraphed the news to Einstein on 27 September 1919. Einstein received the word about this second corroboration of his theory rather placidly. He had been confident about this phenomenon for at least eight years. On 25 October the Dutch Royal Academy held a confirmation meeting in Amsterdam, and on 6 November in

London the Royal Societies there heard Frank Dyson and J. J. Thomson extol Einstein as the worthy successor of Sir Isaac Newton.

Thus suddenly, at the end of 1919 and over the new year 1920, the English Quaker Arthur Eddington and the German pacifist Albert Einstein seemed, as men of science, as idealistic and honorable as Woodrow Wilson or the statesmen at Versailles creating the new League of Nations, which was born in January 1920. Newspapers and magazines everywhere publicized the craze for "relativity," and Einstein's life was never the same again. Because his reputation as the apostle of relativity preceded his character as a theoretical physicist wherever he went, Einstein's technical achievements in synthesizing formerly disparate concepts of "common sense" were seldom appreciated. Rather, he became primarily a figure of hope for a new era. Einstein symbolized a revolution against the tyranny of time.[45]

The power and influence of Einstein's theoretical work on relativity may be recognized in the global scale of the intellectual earthquake that it caused. Even before the returns from the Eddington eclipse expedition were announced, scholars around the world were busy translating and examining the evidence for the restricted and general theories. One of the best and earliest explications of the original papers by Einstein and Minkowski was translated into English by Indian physicists M. N. Saha and S. N. Bose and published at the University of Calcutta in 1920. Perhaps their distance and cultural perspective helped, but their selections, introduced historically by P. C. Mahalanobis, were judicious and honest to the records of the literature. They recognized as clearly and quickly as anyone in Europe the intellectual seismic tremor that the principle of relativity had triggered:

> Einstein's theory connects up the law of gravitation with the laws of motion, and serves to establish a very intimate relationship between matter and physical space-time. Space, time and matter (or energy) were considered to be the three ultimate elements in Physics. The restricted theory fused space-time into one indissoluble whole. The generalized theory has further synthesized space-time and matter into one funda-

mental physical reality. Space, time, and matter taken separately are mere abstractions. Physical reality consists of a synthesis of all three.[46]

On 26 April 1920, at the National Academy of Sciences meeting in Washington, D.C., there occurred another event of major importance for the history of astrophysics. The United States had recently become the new world power in observational astronomy, thanks largely to the entrepreneurship of George Ellery Hale and to the opening of the 100-inch Hooker telescope on Mount Wilson above Los Angeles. Hale staged a dramatic debate over the "scale of the universe" at the Washington meeting by arranging for young Harlow Shapley of Mount Wilson to confront the distinguished Heber D. Curtis of Lick Observatory. The issue was billed by newsmen as one between an expanded galactic universe and the old notion of extragalactic "island universes."

Although both Shapley and Curtis felt comfortable with astrophysics and were intolerant toward mathematicians and metaphysicians, their debate arose over differing interpretations of various kinds of clusters of stars and clouds of light in astronomical photographs. The absolute size of our Milky Way galaxy depended on the question of whether or not spiral nebulae such as Andromeda are galaxies outside of, but similar to, our own "island universe." Curtis, the senior and more conservative advocate, argued broadly for the island-universe hypothesis, explaining spiral nebulae and globular clusters as other galaxies beyond our own. He supposed our sun to be situated nearer the Milky Way's center and our galactic system to be smaller than Shapley suggested. On the other hand, the 34-year-old Shapley, with fresh data from Mount Wilson's 100-inch reflector, challenged this "anthropocentric" view, arguing that our solar system is near the periphery of a much larger swirl of stars than was hitherto thought possible. He tried to show that spiral nebulae, like many open and globular clusters, were truly nebular clouds of stars that are integral parts of the "local" galactic system. Eventually Shapley proved the reality of this "metagalaxy" model and displaced the solar system from its center.[47]

The two protagonists were each half right and half wrong, from the perspective of an additional four decades: Shapley was correct in locating our solar system near the edge of our cluster-filled galactic disk; Curtis was right in judging spirals and some globs as extragalactic galaxies or clusters of galaxies. But neither of the debaters was at this time particularly concerned over the implications of general relativity for gravitation and cosmology. The point is that Heber Curtis and Harlow Shapley represented the tradition of pure, or observational, astronomy. And so rich had been recent progress among disciplined professionals in newly improved observatories that their problems and approaches toward cosmography were virtually independent of Einstein's thus far.

On 5 May 1920, Einstein returned to Leyden to visit Lorentz and the Ehrenfests and to lecture on "Aether and Relativity Theory." In this talk, Einstein presented a compromise on the aether concept that had grown out of his cosmological reconsiderations of the general theory. The infamous "lambda function" (Λ), the cosmical constant he had introduced into the gravitational field equations in 1917 and withdrawn from them in 1919, seems to have revived the problem in the unitary field theory of what to declare superfluous. Einstein's Leyden lecture made a surprising concession, which may have pacified some of the older generation while also placating the revolutionary youngsters. His higher synthesis had led him into avenues of suspicion. Was he now becoming a mere metaphysician?

> Recapitulating, we may say that according to the general theory of relativity space is endowed with physical qualities; in this sense, therefore, there exists an ether. According to the general theory of relativity, space without ether is unthinkable; for in such a space there not only would be no propagation of light, but also no possibility of existence for standards of space and time (measuring-rods and clocks), nor therefore any space-time intervals in the physical sense. But this ether may not be thought of as endowed with the quality characteristic of ponderable media, as consisting of parts which may be tracked through time. The idea of motion may not be applied to it.[48]

Lorentz felt the danger of metaphysics encroaching here, too. Yet, in 1920, he was still arguing, as Einstein well knew, that "It is not necessary to give up entirely even the ether." More concerned than Einstein with the problem of the vacuum in microphysics, Lorentz still conceded that Einstein "put the ether in the background; if he had not done so, he probably would never have come upon the idea that has been the foundation of all his examinations."[49]

While hundreds of books and thousands of articles of all sorts were being published on relativity already, Einstein maintained his balance professionally by keeping one eye on the most profound physical problem of the moment, namely, the relation of quantum theory to elementary electrical phenomena, and the other eye on the promise for extending general relativity into a unified field theory. He must have felt besieged from the left by mathematicians and philosophers, who were eager to exploit the epistemological implications of relativity, and from the right by hordes of patriotic Germans and anti-Semitic protofascists, who with the deepening crises of the infant Weimar Republic, used him as a scapegoat. Two especially bad public meetings in August and September 1920 provoked Einstein into intemperate responses and led him to agree to undertake a trip to the United States the next year in the cause of Zionism.

On 27 January 1921, Einstein delivered before the Prussian Academy in Berlin a discursive address that strongly reaffirmed his position as a theoretical physicist who worried about the mathematicians and philosophers' tendencies to overgeneralize and thus to divorce themselves from experience. The title of this lecture, "Geometry and Experience," asserted his theme, and its thesis was expressed in words that subsequently became widely quoted and plagiarized: "As far as the laws of mathematics refer to reality, they are not certain; and as far as they are certain, they do not refer to reality."[50]

World War I had thoroughly disrupted almost everything in Europe, not least the Swedish ceremonies surrounding the most prestigious honor to which any scientist could aspire, the Nobel Prize. In 1920, there was an effort to catch up with the annual awards, which had been suspended in 1916. Summer ceremonies in Stockholm in June 1920 honored three physicists with retroactive awards for 1917 (C. G. Barkla of England and Scotland "for his discovery of the characteristic Röntgen radiation of the elements"), for 1918 (Max Planck of Berlin "in recognition of the services he rendered to the advancement of Physics by his discovery of energy quanta"), and for 1919 (Johannes Stark of Germany "for his discovery of the Doppler effect in canal rays and the splitting of spectral lines in electric fields"). Later that year, on the usual December day, C. E. Guillaume of Paris received the prize for 1920 for his discovery of anomalies in nickel-steel alloys. In June 1920, Fritz Haber was honored with the 1918 chemistry prize "for his synthesis of ammonia from its elements," and in 1921 Walther Nernst got the 1920 prize for his work in thermochemistry, related to his heat theorem on the third law of thermodynamics.[51]

Then something bizarre and still inexplicable happened to the deliberations among the Nobel Foundation and the Swedish Academy of Science regarding the physics prize for 1921. The committee's decision was delayed a full year and announced only on 10 November 1922, when the prizes for that year were also announced: Frederick Soddy received the 1921 chemistry prize "for his contributions to our knowledge of the chemistry of radioactive substances, and his investigations into the origin and nature of isotopes," and Albert Einstein received the 1921 physics prize "for his services to Theoretical Physics, and especially for his discovery of the law of the photoelectric effect."[52]

The Einsteins had left Europe in October 1922 for a trip to Japan via Suez, Ceylon, and Shanghai. Returning by way of Palestine, Marseilles, and Madrid, they did not arrive home in Berlin until late March 1923. By then, Einstein's antinationalism had created such a stir that the Swedish Ambassador delivered the honors secretly, or at least discreetly, in person. The money, as promised, went to Mileva and Einstein's sons; the medal and parchment, to some mere acquaintance, apparently.[53]

And so it happened that Niels Bohr, recipient of the 1922 Nobel Prize for physics, delivered his Noble Prize lecture on "The Structure of the Atom" some seven months before Einstein arrived in

Sweden to deliver his 1921 Nobel Prize lecture on "The Fundamental Ideas and Problems of the Theory of Relativity." Bohr's long address laid heavy stress on his debts to Einstein and others. It ended with an appeal for intellectual humility in the face of attempts at formal explanations of phenomena too tiny ever to be visualized. Einstein's short address avoided all mention of the photoelectric effect, as if to spite the awards committee, and looked to the present and future rather more than to the past. Einstein stressed the problems of physical relativity (or preferred states of motion in nature) and of operational meanings for all theoretical concepts and distinctions. He recapitulated the special relativity theory, his several syntheses in terms of principles of equivalence, and his general relativity leading into the new theory of gravity. Then he wrestled with Mach's principle once again before confessing difficulty with the "cosmological problem" and announcing the need to embark on a quest for a unified field theory. After giving several credits to Levi-Civita, Weyl, and Eddington for having moved to replace Riemannian geometry with a more general theory of "affine correlation," Einstein concluded his Nobel Prize lecture with the hope that a generalization of the gravitation equations will be found to include the laws of the electromagnetic field in some operationally meaningful way. He was confident that "the relativity principle will [never] be relinquished and the laws previously derived therefrom will at least retain their significance as limiting laws."[54]

Albert Einstein had first met Niels Bohr in 1920, when Bohr went to Berlin to lecture on atomic structure. Each took an immediate liking to the other, and the discussions that ensued continued intermittently for the next thirty-five years. Bohr had owed his original insights of 1912–13 on the quantized structure of the hydrogen atom to Planck, Rutherford, and Einstein. As a foremost spokesman and pioneer in the realm of the microcosmos, Bohr clearly bid fair to become the leader of the next generation of theoretical physicists. And yet his reverence for Einstein was based on the two-thirds of Einstein's work that lay behind and beside his general relativity and cosmology. Einstein recognized and admired Bohr's sure instinct for the essentials, his intuition for the fundamental facts that must be reconciled in theory. Both seemed to know in their bones what could be safely ignored and what deserved worry.

The Quest for a Unified Field Theory

We have come now to the end of the beginning of an intelligent layman's quest for a historical understanding of the genesis of relativity. After Einstein's recognition by his peers about 1910, by the public about 1920, and with *the* prize, acknowledged on 11 July 1923 by his lecture delivered in Göteborg, Sweden, Einstein's greatest achievement was essentially finished. His character and reputation were matched and his major syntheses were, if not fully accepted, being explored for their implications.

But Einstein himself did not recognize this historical judgment. Being only forty-eight years old and in fairly good health, he set forth on the quest that would occupy the next thirty-two years of his life. He considered this search for *the* unified field theory truly a piece of *re*-search, trying to synthesize mathematically the logical foundations for gravitation and electromagnetism. Within the decade he achieved and published *a* unified field theory, but already it had been made obsolete by other advances in physics. And within another decade he would have to account for the strong (and later the weak) "forces" of nucleonic particle interactions. His commitment to the field concept, to continuity, rationality, and causality in nature, was never a blind commitment. But it did become, after the quantum mechanical revolution of the mid-1920s and especially after the fifth Solvay conference, in 1927, a minority position that grew ever less popular. Only after his death was there sizable evidence of a resurgence toward the unification of relativistic and quantum field theories.[55]

To see something of the way in which Einstein moved toward generalizing his general theory, we shall look briefly at a few of the trends and events in observational astrophysics and quantum mechanics that changed the tenor of the times during the 1920s.

The first independent check of the results of the Eddington eclipse expedition to test Einstein's predictions of the bending of starlight was carried out in 1922 by American astronomers R. J. Trumpler and W. W. Campbell, director of the Lick Observatory on Mount Hamilton near San Francisco. Trumpler went to Tahiti in May to prepare instruments and check patterns for comparisons later. On 21 September 1922, at Wallal in Western Australia, W. W. Campbell and his aides succeeded in making good photographs of about seventy-five stars during a total solar eclipse. After data reduction and correlation, they announced their results in April 1923 as being in excellent agreement with Einstein's predictions. Given Campbell's position as an authority on stellar motions and as a former skeptic regarding relativity, his report was especially significant. And it spurred other Californians farther south on to a variety of similar activities.[56]

George Ellery Hale (1868–1938) at the Mount Wilson Observatory was building instruments, institutions, and facilities to enable astronomers of all sorts to probe the limits of man's knowledge. He encouraged many more than Harlow Shapley to challenge accepted doctrines about the universe. Charles E. St. John was one of the first to offer some technical criticism of Einstein's interpretation of the red shift of the solar spectrum. A. A. Michelson made headlines again, in 1920, after adapting his interferometer to the 100-inch Hooker scope to measure stellar diameters. Thereafter, in collaboration with Ludwik Silberstein, Henry G. Gale, and others, Michelson ran several other kinds of experiments also related to relativity testing.[57]

Robert A. Millikan, who at the University of Chicago around 1906 had begun to test Einstein's photoelectric equations, had arrived in 1914, through his oil-drop experiment, at complete confidence in the exact validity of equation and experiment regarding electrons and the photoelectric effect. Millikan moved to Washington, D.C., during World War I to direct the mobilization of U.S. scientists and engineers for the war effort. Then he moved to Pasadena in 1921 and soon became head of the California Institute of Technology. Millikan's early work had become the chief support of both Einstein's law and Bohr's atom, he was honored for this in the presentation speech for his Nobel Prize for physics for 1923. By this time Millikan was also deeply interested in cosmic-ray work.[58]

During this same period, Dayton C. Miller of Case Institute in

Cleveland, the successor to Michelson and Morley there, was invited by George Ellery Hale to finish finally for all seasons of the year and at a 6,000-foot altitude the "infamous" aether-drift tests. Miller's tests on Mount Wilson from 1921 through 1926 were to prove highly irksome and embarrassing to everyone when he announced positive results for an "absolute motion," that is, the resultant of all component motions, for the solar system: about 200 km per second toward the head of the constellation Draco. Although no one else could corroborate Miller's results, Michelson felt challenged and responded in his old age with an elaborate but inconclusive effort to save his "beloved" aether. Lorentz and Einstein also became involved in Miller's challenge.[59]

Meanwhile, Edwin P. Hubble and Milton L. Humason at Mount Wilson were fascinated by the problems raised by Einstein and St. John. Through most of the 1920s, they sought to check the points where positional astronomy, spectroscopy, and galactic structure could be juxtaposed.

Radial motions of stars and galaxies at the "rim" of the universe became measurable about this time. The *proper motions* (motions *across* the line of sight) of so-called fixed stars had become a major astronomical concern in the late nineteenth century, but *radial motions* (motions *along* the line of sight) only became feasible after the pioneering work of such people as J. C. Kapteyn (1851–1922) on star streaming, E. E. Barnard (1857–1923) on stellar photography, Lewis Boss (1846–1912) and his son Benjamin on cataloging, and Vesta M. Slipher (1875–1969) on recessional velocities. Around 1920, the exciting endeavor to apply the Doppler principle to stellar spectra in order to find red or blue shifts indicating the recessions or approaches of distant celestial objects was led by W. W. Campbell at Lick and in Chile, by A. A. Belopolski (1854–1934) at Pulkova in Russia, and by Frank Schlesinger (1871–1943) and William H. Wright (1871–1959) in the United States.[60]

In 1924, Hubble announced new evidence that many fuzzy nebulae and some globular clusters were indeed distant galaxies far beyond our own. This effectively settled the great debate over the scale of the sidereal universe in favor of extra-galactic nebulae. But the continuing spectral analyses for radial velocities by Humason at Mount Wilson and others elsewhere led into the concept of Shapley's "metagalaxy," which could for many purposes allow our own Milky Way stellar system to be considered, as Newton had considered our solar system, isolated, unique, and virtually universal.

All this excitement and new knowledge issuing from Mount Wilson's 100-inch eye led Hale to plan for a 200-inch telescope. In 1928, he obtained Carnegie money and started a publicity campaign that eventuated in the building of the Palomar Observatory and the dedication in 1948 of the 200-inch Hale telescope. Meanwhile and worldwide, astronomy was advancing.[61]

Hubble's law of 1929 correlating the velocity-distance relation as a linear function quickly became linked with the apparently uniform distribution of the nebulae outside our galaxy. These two fundamental principles derived from observational astronomy became the new bases upon which all cosmological theories had to rest. Hubble's *Realm of the Nebulae* (1936) seemed to endorse the homogeneity of matter in space, and Einstein's general relativity theory had postulated this as well as the isotropy of its geometrical structure.[62]

Willem de Sitter had first challenged Einstein's use of lambda, his cosmological constant, in 1917, arguing for an empty yet expanding, rather than a full yet static, model of the universe. The Russian scientist Aleksandr A. Friedmann had likewise argued in 1922 and 1924 for both expansion and contraction of various "island universes." Such local oscillations or total pulsations seemed to fit the growing mass of observational data by about 1930. Also by then the Belgian priest Abbé Georges Lemaître had hit upon his thermodynamically supported hypothesis for a "primeval atom." This notion, scientifically elaborated through 1950, provided a way of reconciling the Judeo-Christian mythos of creation with the most modern theories of cosmogony. This taproot for what became the so-called big-bang thesis was enthusiastically endorsed by Eddington and gradually accommodated by Einstein, who came to think of his lambda constant as his biggest blunder.[63]

A leading student of the intricacies of these cosmological models was George Gamow (a student of Friedmann as well as of Bohr and Born), who left Soviet Russia in 1933. While doing much serious work on cosmic radiation and on helium synthesis, Gamow also became a leading popularizer of "big bang" cosmology. In contrast there arose by 1948 a group of "steady state" theorists, led by Hermann Bondi, Thomas Gold, and Fred Hoyle, who insisted on a "perfect cosmological principle" in order to keep cosmology scientific. By this they meant that the universe as a whole must be considered homogeneous and isotropic, therefore as being continuously created and destroyed in a balanced equilibrium. By 1940, the work of R. W. Fowler, Carl F. von Weizsäcker, Hans A. Bethe, and Gamow, among others, had led to the suggestion that "cosmochemistry" be considered coordinate with the cosmology that had grown out of the geometrical and mechanical emphases of Einstein's general theory. By that time general relativity and quantum theory were complementary in practice, just as astrophysics and cosmochemistry were to become.[64]

Meanwhile, throughout the 1920s, Einstein had continued to struggle with his quest for a unitary field theory. After the rather shocking reports in 1923 from Arthur H. Compton at St. Louis that he had been able to produce experimental proof that X rays can be *totally* reflected from glass or silver mirrors at small angles of incidence *and* that X rays experience a change in wave length, a "softening," in scattering experiments, physicists had to cope with this strong direct evidence that X rays show *both* a wave and a particle nature. Thus the dilemma of wave-particle duality, reinforced thereafter by similar news from elsewhere, led physicists much more deeply into the quantum interpretations of Planck and Einstein regarding radiant energy transfer. In Paris, the younger brother of Maurice de Broglie, Prince Louis de Broglie, was finishing his doctoral thesis and publishing during 1923–24 a series of papers showing how matter (in the form of electrons and protons) and radiation (in the form of quanta or "atoms of light") might be reconciled. De Broglie conceived every material particle to show a "phase wave" and every "ray" to carry discrete resonance packets similar to such "material" particles as electrons but much smaller. Just as Einstein had found particulate properties in radiation, so now de Broglie found wave properties in particles of matter.[65]

Einstein received the evidence of the Compton effect and the ideas of de Broglie with interest. During this period he was collaborating with Paul Ehrenfest and Hans Mühsam on different aspects of quantum theory and with Jacob Grommer in criticizing the five-dimensional field theory of T. Kaluza. While working also on a quantum approach to monatomic ideal gases, Einstein received a paper entitled "Planck's Law and the Hypothesis of Light Quanta" from S. N. Bose of the University of Dacca in India; this so impressed him that he translated it for publication in German. Thus began the Bose-Einstein statistics, which, when applied to aggregates of light-quanta, enabled one to treat massless particles as if they were molecules of a degenerating ideal gas. Such "bosons," as they became known several decades later in contrast to "fermions" (named after heavier massive-particles that would obey Fermi-Dirac statistics), were appropriately named after Bose.[66]

After their Nobel prizes had been awarded, Einstein and Bohr began to notice certain subtle differences in their taste and style of doing physics. This divergence ripened through the crisis that was approaching. The best of Max Born's students at Göttingen— Wolfgang Pauli and Werner Heisenberg—were visiting Bohr's institute in Copenhagen more than Einstein's institute in Berlin. It seemed as if the problem of atomic structure was more compelling, if not more tractable, than that of the unified field. Theoretical spectroscopy based on laboratory results seemed more promising than that based on observatory data. And yet Einstein retained a foot in both camps, as was recognized by physicists like Alfred Landé and Erwin Schrödinger.[67]

In June and July of 1925, the first fruits from the divergent but prepared minds of Heisenberg and Schrödinger began to appear. These works were destined within a year or two to transform the old quantum theory into the new quantum mechanics. By almost antithetical approaches Heisenberg, using Bohr's correspondence principle, and Schrödinger, using Einstein's wave-particle duality, produced almost simultaneously new mathematical techniques to resolve the paradoxes of matter and radiation theory as then acknowledged. Heisenberg, with the aid of Max Born and Pascual Jordan, began to develop a noncommutative multiplication rule into the analytical tool called matrix mechanics. They applied matrix algebra to the spectra of each family of the chemical elements in the Periodic Table and found, with the help of Bohr, Pauli, and others, that they had broken the spectral code. The atomic structure and characteristic spectra of each element promised to yield their secrets. [68]

In almost perfect parallel, Schrödinger, inspired by Broglie and Einstein, began to develop a wave function from fluid mechanics into the analytical tool called wave mechanics. He applied the Bose-Einstein statistical procedures together with Broglie's matter waves to the Planck radiation law, thence to Einstein's new theory of gas degeneracy in order to form a picture "taking seriously the de Broglie-Einstein wave theory of moving particles, according to which the particles are nothing more than a kind of "wave crest" on a background of waves."[69]

Despite the superficial similarities and profound differences between Heisenberg's and Schrödinger's systems for understanding microphysics, yet another system came forth during 1926 to complement matrix and wave mechanics and to translate the whole of classical dynamics into quantum theory. This brilliant achievement of Paul A. M. Dirac of Cambridge University brought a third scheme of explanation into play. And thus the revolution characterized by three equally powerful mathematical approaches to matter and radiation, known collectively as quantum mechanics, proved irresistible.[70]

In 1927, theory, it seemed, had finally caught up with experimentation. And yet at the fifth Solvay conference in Brussels, that year, devoted to the theme "Electrons and Photons," the senior theoretical physicists found themselves profoundly polarized by conflicting philosophies. Bohr came to explain his schools' working consensus that operational meanings alone should be trusted, that probabilities within the various forms of quantum statistics should replace the search for causal laws, and that all observers should recognize henceforth that the act of observation may change or otherwise affect what is observed. Heisenberg's new principle of indeterminacy, stating that "the product of the uncertainties in the measured values of the position and momentum [that is, the product of mass and velocity] cannot be smaller than Planck's constant," had just been coined.[71] It served Bohr well as a central feature of his Copenhagen interpretation of new bases for physics. Just a month before, in September 1927 at Lake Como in Italy, Bohr had launched his principle of complementarity to another congress of physicists: "Evidence obtained under different experimental conditions cannot be comprehended within a single picture, but must be regarded as *complementary* in the sense that only the totality of the phenomena exhausts the possible information about the objects."[72]

In Brussels, Einstein arose in response to this challenge, arguing forcefully against such trust placed in discontinuities, probabilities, indeterminacy, and complementarity. All his instincts by now inveighed against accepting the Dane's doctrines. To Einstein it seemed puerile to think that physics had reached a state of perfection or completeness. To Bohr it seemed equally futile or childish to believe that men will ever become divine enough to know everything in detail about either universal or individual events. Bohr and his allies felt that quantum-mechanical descriptions would eventually and essentially exhaust the possibilities of accounting for observable phenomena. Einstein vehemently disagreed, lamenting the loss to science of causality, continuity, and determinism.[73]

Three years later, at the sixth Solvay conference, the Einstein-Bohr debate continued even more dramatically. The year before, in 1929, Einstein had with great hope published a unified field theory; it was poorly received, probably because quantum mechanics and atomic and nuclear physics promised so much more in return. Then, too, all the developments from quantum theory fed back into electromagnetism and thus modified half of what Einstein had unified. At any rate, Einstein confronted Bohr in 1930 with a new thought-experiment that attempted to prove the principle of complementarity was untenable. To Einstein, Bohr had made the mistake of permitting theoretical description to be too directly dependent on mere observations or experiments. But Bohr defended himself heartily, and after he showed Einstein had forgotten certain implications of his own general theory of relativity, Einstein again had to retreat. Each felt that the other was verging on a betrayal of the spirit of physics.[74]

By 1930, the leaders of the profession could see an obvious epistemological split between the rationalism of Einstein and the relativism of Bohr. Ironically, as the years passed, Einstein became less a leader of revolutionary relativism, a critic of accepted concepts, and a creator of startling new syntheses, and more a conservative spokesman for what the younger generation thought of as "classical" physics. He was still deeply immersed in his quest for a unified field theory, but he recognized his growing loneliness in that search. With wry humor he tried to keep up with the avalanching advances in physical discovery while maintaining as a personal program the research for a set of continuous functions in the four-dimensional space-time continuum. (He could afford this pursuit of a possible chimera; younger men could not.) The basic obstacle to achieving a larger field theory for the synthesis of gravitation and electromagnetism was the perpetual appearance of "singularities," that is, places or parts of space-time where the field equations are not valid. In his efforts to avoid such singularities, Einstein was led farther from science and deeper into philosophy.[75]

NOTES

1. Gerald Holton, *Thematic Origins of Scientific Thought: Kepler to Einstein* (Cambridge: Harvard Univ. Press, 1973), esp. pp. 240–56 and article 10 "On Trying to Understand Scientific Genius" from 1971, pp. 353–80. See also Jacques Merleau-Ponty and Bruno Morando, *The Rebirth of Cosmology*, trans. Helen Weaver (New York: Alfred A. Knopf, 1976).

2. Banesh Hoffmann with Helen Dukas, *Albert Einstein: Creator and Rebel* (New York: Viking Press, 1972), p. 3. L. Pearce Williams, ed., *Relativity Theory: Its Origins and Impact on Modern Thought* (New York: Wiley, 1968), pp. 115–57.

3. Albert Einstein, "Autobiographical Notes," in Paul A. Schilpp, ed., *Albert Einstein: Philosopher-Scientist* (New York: Harper Torchbooks, 1949), pp. 5, 7, 17.

4. A. Einstein, "Über die Entwicklung unserer Anschauungen über das Wesen und die Konstitution der Strahlung," *Physikalische Zeitschrift* 10 (1909): 817–29.

5. Probably the best account of "Einstein at Prague" is that written by his successor there, Philipp Frank, *Einstein: His Life and Times*, trans. George Rosen (New York: Knopf, 1967), chap. 4, pp. 74–100, orig. German ed., Munich, 1949.

6. On the Solvay conferences, see Jagdish Mehra, *The Solvay Conferences on Physics: Aspects of the Development of Physics Since 1911* (Dordrecht: D. Reidel, 1975). A. Einstein, "On the Influence of Gravitation on the Propagation of Light," from *Annalen der Physik* (4) 35 (1911): 898–908 in translation by W. Perrett and G. B. Jeffery, *The Principle of Relativity: A Collection of Original Memoirs . . . by H. A. Lorentz, A. Einstein, H. Min-*

kowski and H. Weyl (London: Methuen, 1923 and Dover reprint) will be used here. Quotation is from p. 130.

7. On the question of Einstein's citizenship, see Carl Seelig, *Einstein: A Documentary Biography* (London: Staples, 1956).

8. For the texts of these manifestoes and their context, see Otto Nathan and Heinz Norden, eds., *Einstein on Peace* (New York: Schocken Books, 1968), pp. 1–26, original edition 1960. For Einstein's friendship with Ehrenfest, see Martin J. Klein, *Paul Ehrenfest: The Making of a Theoretical Physicist* (New York: American Elsevier, 1970), vol. I.

9. Armin Hermann, ed., *Albert Einstein/Arnold Sommerfeld: Briefwechsel* . . . (Basel-Stuttgart: Schwabe, 1968), pp. 32–36. Albert Einstein, "The Origins of the General Theory of Relativity," pamphlet of First Gibson Foundation Lecture, 20 June 1933 (Glasgow: University Publications XXX, 1933), p. 8.

10. A. Einstein, "Die Grundlage der Allgemeinen Relativitätstheorie," *Annalen der Physik,* 4th series, 49, No. 7 (1916): 769–822; and in a 64-page pamphlet under same title (Leipzig: J. A. Barth, 1916). The standard English translation by Perrett and Jeffery in *The Principle of Relativity.*

11. A. Einstein, *The Principle of Relativity,* p. 113. Some English spellings have been Americanized in the following quotations. Italics are original.

12. Ibid., p. 117.

13. Ibid. p. 143.

14. Ibid., pp. 148, 149.

15. Ibid., p. 152.

16. Ibid., p. 161.

17. Ibid., pp. 162, 163. This third test was prefabricated: "Calculation gives for the planet Mercury a rotation of the orbit of 43" per century, corresponding exactly to astronomical observation . . ." (p. 164).

18. A. Einstein, "Ist die Trägheit eines Körpers von seinem Energiegehalt abhängig?" *Annalen der Physik* (4) 18 (1905): 639–41, in the Perrett and Jeffery translation, op. cit., pp. 69–71.

19. This section is based largely on Max Jammer, *Concepts of Mass in Classical and Modern Physics* (Cambridge: Harvard Univ. Press, 1961). Harper Torchbook edition, 1964, esp. pp. 136–224.

20. Ibid., p. 153.

21. Ibid., pp. 177–80.

22. See, e.g., Gustave Le Bon, *The Evolution of Forces* (New York: Appleton, 1908) and Joseph Petzoldt, ed., *Zeitschrift für positivistische Philosophie* I, No. 1, November 1912.

23. Jammer, *Concepts of Mass,* pp. 184–88.

24. For an authorized review of Wheeler's thought in context, see John C. Graves, *The Conceptual Foundations of Contemporary Relativity Theory* (Cambridge: MIT Press, 1971).

25. See chap. IV, pp. 178–79, above. Helpful throughout this section is Max Jammer, *Concepts of Space: The History of Theories of Space in Physics* (Cambridge: Harvard Univ. Press, 1954). Harper Torchbook edition, 1960, with a foreword by Albert Einstein. See also Peter A. Bowman, "Einstein's Second Treatment of Simultaneity, *PSA 1976* (Philosophy of Science Association) 1 (1976): 71–81.

26. This and the next section owe much to Jagdish Mehra, *Einstein, Hilbert and the Theory of Gravitation: Historical Origins of General Relativity Theory* (Dordrecht: D. Reidel, 1974); but, regarding the Hilbert-Einstein relationship, see also the forthcoming works by John Earmon and Clark Glymour.

27. For H. Minkowski's "Raum und Zeit" address, see the Perrett and Jeffery translation with notes by Arnold Sommerfeld in *The Principle of Relativity*, pp. 75–96; quotation from p. 80.

28. Mehra, *Einstein, Hilbert*, pp. 2–16; cf. Constance Reid, *Hilbert* (New York: Springer-Verlag, 1970), Chap. 16, "Physics."

29. Cf. Carl Boyer, *A History of Mathematics* (New York: Wiley, 1968), pp. 652–62, and Morris Kline, *Mathematical Thought from Ancient to Modern Times* (New York: Oxford Univ. Press, 1972) pp. 1192–1208.

30. Cf. G. J. Whitrow, *The Natural Philosophy of Time* (London: Nelson & Sons, 1961). Harper Torchbook edn., 1963, pp. 176–267; and Whitrow, *Time, Gravitation and the Universe: The Evolution of Relativistic Theories* (London: Imperial College of Science and Technology, 1973); Inaugural Lecture, 22 May 1973.

31. See also Einstein and Grossmann, "Kovarianzeigenschaften der Feldgleichungen der auf die verallgemeinerte Relativitätstheorie gegründeten Gravitationtheorie," *Zeitschrift für Mathematik und Physik* 63 (1914): 215–25. R. H. Dicke, "The Eötvös Experiment," *Scientific American* 205 (December 1961): 84–94.

32. On Einstein's estrangement from his first family, see the testimony of his oldest son, Hans A. Einstein, in G. J. Whitrow, ed., *Einstein: The Man and His Achievement* (New York: Dover, 1973), pp. 18–22. Re Einstein's memory of this period, see his Gibson Lecture at Glasgow in 1933, cited in note 9. See also two essays, C. Lanczos, "Einstein's Path from Special to General Relativity," and N. L. Balazs, "The Acceptability of Physical Theories: Poincaré vs. Einstein," in L. O'Raifeartaigh, ed., *General Relativity: Essays in Honor of J. L. Synge* (Oxford: Clarendon Press, 1972).

33. Mehra, *Einstein, Hilbert*, pp. 17–50.

34. Ibid., p. 44.

35. Max Jammer, *Concepts of Force: A Study in the Foundation of Dynamics* (Cambridge: Harvard Univ. Press, 1957). Harper Torchbook edn., 1962, pp. 200–264.

36. See Eugene Guth, "Contribution to the History of Einstein's Geometry as a Branch of Physics," in Moshe Carmeli, et al., eds., *Relativity: Pro-*

ceedings of a Conference . . . at Cincinnati, Ohio, June 1969 (New York: Plenum Press, 1970), pp. 161–200.

37. For some insight into the Adler affair, see Lewis S. Feuer, *Einstein and the Generations of Science* (New York: Basic Books, 1974), pp. 14–26. While imprisoned, Friedrich Adler wrote a remarkable book comparing the attitudes toward time of Woldemar Voigt, H. A. Lorentz, and A. Einstein: *Ortszeit, Systemzeit, Zonenzeit . . .* (Vienna: Wiener Volksbuchhandburg, 1920).

38. Max Wertheimer, *Productive Thinking,* enlarged ed. 1959, Michael Wertheimer, ed. (New York: Harper, 1959), chap. 10, pp. 213–33: "Einstein: The Thinking That Led to the Theory of Relativity." For a thorough contextual analysis, see Arthur I. Miller, "Albert Einstein and Max Wertheimer: A Gestalt Psychologist's View of the Origins of Special Relativity Theory, *History of Science* 13 (1975): 75–103.

39. A. Einstein, "Quantentheorie der Strahlung," *Physikalische Zeitschrift* 18 (1917: 121–28. See Sir James Jeans's section on "Einstein's Synthesis" in his *Physics and Philosophy* (Cambridge: Univ. Press, 1944), pp. 150–51.

40. A. Einstein, "Cosmological Considerations on the General Theory of Relativity" (1917) in the Perrett and Jeffery translation, pp. 177–88, esp. pp. 178, 180, 182, 183, 186, and 188. For an excellent overview see John David North, *The Measure of the Universe: A History of Modern Cosmology* (Oxford: Clarendon Press, 1965).

41. Hermann Weyl, *Raum, -Zeit, -Materie* (Berlin, 1918); 4th edn., 1921, trans. Henry L. Brose, *Space-Time-Matter* (New York: Dover, 1922, 1950). The influence of Weyl was both immediate and long-range with his "gauge" invariance and unified field attempts. The translator, Brose, was also author of *The Theory of Relativity: An Introductory Sketch* (Oxford: Blackwell, 1919).

42. A. Einstein, "Do Gravitational Fields Play an Essential Part in the Structure of the Elementary Particles of Matter?" (1919) in the Perrett and Jeffery translation, pp. 191–98, esp. pp. 192 and 198. Notation changes for the tensors and subscripts (such as $G_{\mu\nu} \rightleftharpoons R_{\iota\kappa}$) in later years left the form of Einstein's field equation intact, but its content remained problematic. In his 1936 essay, "Physics and Reality," Einstein wrote of it as follows:

"By this formulation one reduces the whole mechanics of gravitation to the solution of a single system of covariant partial differential equations. The theory avoids all internal discrepancies which we have charged against the basis of classical mechanics. It is sufficient—as far as we know—for the representation of the observed facts of celestial mechanics. But, it is similar to a building, one wing of which is made of fine marble (left part of the equation), but the other wing of which is built of low-grade wood (right side of equation). The phenomenological representation of matter is, in fact, only a crude substitute for a representation which would correspond to all known properties of matter."

See Albert Einstein, *Out of My Later Years*, rev. edn. (Westport, Conn.: Greenwood Press, 1970), p. 83.

43. For an illustrated biography, see Banesh Hoffmann, with Helen Dukas, *Albert Einstein: Creator and Rebel* (New York: Viking Press, 1972), esp. pp. 134 ff. See also William Cahn, *Einstein: A Pictorial Biography* (New York: Citadel Press, 1955).

44. Sir Arthur S. Eddington, *Space, Time and Gravitation: An Outline of the General Relativity Theory* (London, 1920). Harper Torchbook edn. (1959), pp. 113–22; cf. Eddington's *Stellar Movements and the Structure of the Universe* (London: Macmillan, 1914), pp. 54, 71 ff. See also Clark, *Einstein: The Life and Times*, op. cit., p. 229, for Eddington's poetic report.

45. J. J. Thomson's hosanna about Einstein's relativity: "one of the greatest—perhaps the greatest—of achievements in the history of human thought" was perhaps typical of the exultant mood. For Continental reactions, see the biography by a colleague in literary circles in Berlin, Alexander Moszkowski, *Einstein the Searcher*, trans. Henry L. Brose (London: Methuen, 1921).

46. P. C. Mahalanobis, "Historical Introduction" to anthology translated by M. N. Saha and S. N. Bose, *The Principle of Relativity: Original Papers by A. Einstein and H. Minkowski* (Calcutta: Univ. of Calcutta Press, 1920), p. xxii. For contrasting ahistorical attitudes of most Western students, see the various essays competing for the $5000 Higgins prize, in James Malcolm Bird, ed., *Einstein's Theories of Relativity and Gravitation* (New York: Scientific American Publishing, 1921).

47. For the great debate on "The Scale of the Universe," see *Bulletin of the National Research Council*, vol. II, pt. 3, no. 11 (Washington, D.C., 1921). See also Otto Struve, *The Universe* (Cambridge: MIT Press, 1962), pp. 71–76. Willy Ley, *Watchers of the Skies: An Informal History of Astronomy* . . . (New York: Viking Press, 1963), Compass edn. (1968), pp. 471–77. Cf. N. S. Hetherington, "The Shapley-Curtis Debate," *Astronomical Society of Pacific*, Leaflet no. 490 (April 1970). For an excellent illustrated account, see Richard Berendzen, Richard Hart, and Daniel Seeley, *Man Discovers the Galaxies* (New York: Science History Publications, 1976). And especially, see Harlow Shapley's autobiography, *Through Rugged Ways to the Stars* (New York: Scribner's, 1969), pp. 75–85.

48. Albert Einstein, "Ether and Relativity Theory," address delivered 5 May 1920 at the University of Leyden, trans. G. B. Jeffery and W. Perrett, in *Sidelights on Relativity* (London: Methuen, 1922), pp. 23–24.

49. H. A. Lorentz, *The Einstein Theory of Relativity: A Concise Statement* (New York: Brentano, 1920), pp. 60, 63. Despite his admiration for

Einstein, J. J. Thomson felt like Lorentz toward the aether and the vacuum: see George P. Thomson, *J. J. Thomson: And the Cavendish Laboratory in His Day* (Garden City, N.Y.: Doubleday, 1965), p. 155 and "Appendix on Vacuum Pumps," pp. 176–81. See also G. W. C. Kaye, *High Vacua* (London: Longmans, 1927).

50. A. Einstein, "Geometry and Experience" in Jeffery and Perrett translation from *Sidelights of Relativity*, p. 28, and most conveniently reprinted in *Ideas and Opinions by Albert Einstein* (New York: Bonanza Books, 1954), p. 233.

51. Nobel Foundation, *Nobel Lectures, Chemistry, 1901–1921* (Amsterdam: Elsevier, 1966). Nobel Foundation, *Nobel Lectures, Physics, 1901–1921* Amsterdam: Elsevier, 1967). Nobel Foundation, *Nobel Lectures, Physics, 1922–1941* (Amsterdam: Elsevier, 1965).

52. Re Einstein's Nobel Prize award, see Jeremy Bernstein, *Einstein* (New York: Viking Press, 1973), pp. 187–90. See also Harriet Zuckerman, "The Sociology of Nobel Prizes," *Scientific American* 217 (November 1967): 25–33, and her book *Scientific Elite: Nobel Laureates in the United States* (New York: Free Press, 1977). Strange as was this episode, it was hardly more peculiar than the life of the bequestor himself; see Nicholas Halasz, *Nobel: A Biography of Alfred Nobel* (New York: Orion Press, 1959).

53. The best family biography is by "Anton Reiser," really Rudolf Kayer, a stepson-in-law (Ilse's husband): *Albert Einstein: A Biographical Portrait* (New York: A. & C. Boni, 1930).

54. Cf. Niels Bohr, *Nobel Lectures, Physics, 1922–1941*, pp. 1–47. Albert Einstein, *Nobel Lextures, Physics, 1901–1921*, pp. 479–92.

55. Einstein's scientific thought after 1920 may be conveniently traced by reading the essays chronologically arranged by Valentine and Sonja Bargmann in *Ideas and Opinions by Albert Einstein*, pp. 217–377. For an interesting comparative study by another physician, see Aaron B. Lerner, *Einstein and Newton: A Comparison of the Two Greatest Scientists* (Minneapolis: Lerner Publications, 1973).

56. See the Campbell-Trumpler report of 1923 in Harlow Shapley, ed., *Source Book in Astronomy, 1900–1950* (Cambridge: Harvard Univ. Press, 1960), pp. 351–55 (hereafter cited as Shapley, *Source Book II*). See also W. W. Campbell, *Stellar Motions* (New Haven: Yale Univ. Press, 1913).

57. For Hale, see Helen Wright, *Explorer of the Universe: A Biography of George Ellery Hale* (New York: Dutton, 1966), and Helen Wright, Joan N. Warnow, and Charles Weiner, eds., *The Legacy of George Ellery Hale: Evolution of Astronomy and Scientific Institutions in Picture and Documents* (Cambridge: MIT Press, 1972). For Michelson et al. see Loyd S. Swenson, *The Ethereal Aether: A History of the Michelson-Morley-Miller Aether-Drift Experiments, 1880–1930* (Austin: Univ. of Texas Press,

1972), pp. 190–212. See also the excellent overview by Daniel J. Kevles, *The Physicists: The History of a Scientific Community in Modern America* (New York: Alfred A. Knopf, 1978).

58. Robert A. Millikan, "The Electron and the Light-Quant from the Experimental Point of View," Nobel lecture, 23 May 1924, in *Nobel Lectures, Physics, 1922–1941*, pp. 54–69. See also his books, *Electrons (+ and −), Protons, Photons, Neutrons, and Cosmic Rays* (Chicago: Univ. of Chicago Press, 1935) and *Autobiography* (New York: Prentice-Hall, 1950).

59. See Swenson, *The Ethereal Aether*, op. cit., chap. 10, "Miller Challenges Michelson, 1920–1925," and chap. 11, "Michelson Reaffirms the Null, 1925–1930." See also Dayton C. Miller, "The Ether-Drift Experiment and the Determination of the Absolute Motion of the Earth," *Reviews of Modern Physics* 5 (July 1933): 203–41.

60. Shapley's *Source Book II* devotes sec. IV to "The Positions and Motions of the Stars," sec. XI to "Galaxies," and sec. XII to "Relativity and Cosmogony." See esp. pp. 103–4, 124–29, 140–46. See also Thornton and Lou Williams Page, eds., *Stars and Clouds of the Milky Way: The Structure and Motion of our Galaxy* (New York: Macmillan, 1968).

61. Cf. Georgio Abetti, *The History of Astronomy*, trans. B. B. Abetti (London: Abelard-Schuman, 1952), orig. edn. Rome, 1951; and A. Pannekoek, *A History of Astronomy*, (New York: Interscience, 1961), orig. edn. Amsterdam, 1951. See also James S. Hey, *The Evolution of Radio Astronomy* (New York: Neale Watson Academic, 1973).

62. Edwin Hubble, *The Realm of the Nebulae* (New Haven: Yale Univ. Press, 1936); Dover reprint, 1958; and Hubble, *The Observational Approach to Cosmology* (Oxford: Clarendon Press, 1937). See also H. P. Robertson, "Geometry as a Branch of Physics," in P. A. Schilpp, ed., *Albert Einstein: Philosopher-Scientist*, vol. I., pp. 315–32. For a final tribute, see Allan Sandage, *The Hubble Atlas of Galaxies* (Washington, D.C.: Carnegie Institution, 1961).

63. Willem de Sitter, *Kosmos* [the Lowell lectures, Boston, November 1931] (Cambridge: Harvard Univ. Press, 1932). Canon Georges Lemaître, *The Primeval Atom: An Essay on Cosmogony* (New York: Van Nostrand, 1950), trans. B. H. and S. A. Korff. Cf. Shapley, *Source Book II*, article by Lemaître, pp. 374–81.

64. George Gamow, *My World Line: An Informal Autobiography* (New York: Viking Press, 1970), pp. 41–45, 94–99. See also Shapley, *Source Book II*, article by Bondi and Gold (1948), pp. 382–85, article by Rupert Wildt (1940), pp. 394–401, and article by Otto Struve (1948), pp. 402–10.

65. Roger H. Stuewer, *The Compton Effect: Turning Point in Physics* (New York: Science History Publications, 1975). Martin J. Klein, "Einstein and the Wave-Particle Duality," in Gershenson-Greenberg, eds., *The Natural Philosopher*, vol. I, pp. 3–49, and Bobbs-Merrill reprint HS-37 (1964).

66. On Fermi, see Gerald Holton, "Striking Gold in Science. Fermi's

Group and the Recapture of Italy's Place in Physics," *Minerva* 12 (April 1974): 159–98. In addition to S. N. Bose's new fame in the West, his collaborator in the 1920 translations, Megh Nad Saha, had also achieved eminence in 1921 with a new theory of stellar spectra; see five references in Shapley, *Source Book II*.

67. For Bohr's memoirs, see Niels Bohr, "The Genesis of Quantum Mechanics" (1962), in his *Essays 1958–1962 on Atomic Physics and Human Knowledge* (New York: Vintage, 1966), pp. 74–78. See also Werner Heisenberg, *Physics and Beyond: Encounters and Conversations*, trans. Arnold J. Pomerans (New York: Harper & Row, 1971), pp. 43–69.

68. Serious study of the quantum mechanics revolution must begin with Thomas S. Kuhn, John L. Heilbron, Paul Forman, and Lini Allen, *Sources for History of Quantum Physics: An Inventory and Report* (Philadelphia: American Philosophical Society, 1967).

69. For Schrödinger's 15 December 1925 paper "On Einstein's Gas Theory" in translation (excerpts), see Martin J. Klien, "Einstein and the Wave-Particle Duality," p. 43. See also V. V. Raman and Paul Forman, "Why Was It Schrödinger Who Developed de Broglie's Ideas?" in Russell McCormmach, ed., *Historical Studies in the Physical Sciences*, I (1969): 291–314.

70. See Max Jammer, *The Conceptual Development of Quantum Mechanics* (New York: McGraw-Hill, 1966). For a reliable, readable popularization, see Barbara B. Cline, *The Questioners: Physicists and the Quantum Theory* (New York: Crowell, 1965), pp. 179–91. See also Karl Przibaum, ed., *Letters on Wave Mechanics: Schrödinger, Planck, Einstein, Lorentz*, trans. Martin J. Klein (New York: Philosophical Library, 1967).

71. The name "photons" for light-quanta had also just been coined by Gilbert N. Lewis in "The Conservation of Photons," *Nature* 118 (18 December 1926): 874–75. For Heisenberg's recount of his uncertainty principle, see *Physics and Beyond*, p. 78.

72. Niels Bohr, "Discussion with Einstein on Epistemological Problems in Atomic Physics," in P. A. Schilpp, ed., *Albert Einstein: Philosopher-Scientist*, pp. 209–10. This whole essay, plus Einstein's reply (pp. 666–76) to Bohr et al, is basic to what follows.

73. For a masterful probe of "The Roots of Complementarity," see Gerald Holton, *Thematic Origins of Scientific Thought*, pp. 115–61.

74. A. Einstein, "Zur Einheitlichen Feldtheorie . . ." (Berlin: W. de Gruyter, 1929) and explicated in English in "Field Theories, Old and New," pamphlet reprint of article in *New York Times*, 3 February 1929. See also Lewis S. Feuer, *Einstein and the Generations of Science*, op. cit. pp. 109–36.

75. The continuation of the Bohr-Einstein debate centered on the "completeness" argument of 1935 may be traced through Stephen Toulmin, ed., *Physical Reality: Philosophical Essays on Twentieth Century Physics* (New York: Harper Torchbooks, 1970).

Epilogue on Einstein

"Relativity," as a concept, is a characteristic feature of the twentieth century. It has become much more than the restricted and general theories of relativity that developed out of work by Albert Einstein. And Einstein no more invented the theory of relativity than Darwin invented the theory of evolution. Just as the idea of evolution was in the air in mid–nineteenth century Europe, so also the idea of relativity was a pervasive notion by the beginning of the twentieth century. As we have seen, Einstein himself admitted as much regarding the special theory, but he was much more ego-involved in its generalization into his theory of gravitation. And yet honest historians must admit that the state of mathematical physics had reached a level where any one of a dozen or so of his peers might, under slightly different circumstances, have become known as the "father" of the relativity revolution. No one else, however, had his unique combination of genes, talents, experiences, and persistence. And no one else carried the search for unity in nature quite so far so successfully. Only Albert Einstein sought and found his personal identity in a *series* of theories of relativity.

The historical genesis of the unique relativity theory that we have traced through these pages represents the advent of a single mind thinking hard within a discipline about the problems and paradoxes, often hidden from the view of others, that confronted those people who were probing the limits of human knowledge. Downward into the microcosmos and outward into the macrocosmos, from the realm of nuclei to that of nebulae, science as the advancing edge of objectivity, with the aid of instruments for probing, testing, sensing, and measuring the world around us, extended human awareness beyond belief.[1]

Anthropologists were discovering that each person is in some respects like all other people, in many respects like most other people, and in a few respects like no other person. Cultural relativity was growing to pervade the social sciences well before most of the priests of the physical sciences came to appreciate the fact that science must take account of subjects as well as objects, of observers as well as the observed, of private as well as public aspects of knowledge.[2]

Einstein was preternaturally sensitive to these trends at first. But with age his quest for the unified field, together with his super-professionalization and exile and isolation, led him to deny Bohr's principle of complementarity and Heisenberg's principle of uncertainty. The so-called Copenhagen interpretation became anathema to him after 1927, when he felt himself near the third and culminating stage of his development of the theory of relativity. In 1929, Einstein was pleased to explain himself in a newspaper article, "Field Theories, Old and New," after having just finished unifying the field laws for gravitation and electromagnetism:

The characteristics which especially distinguish the general theory of relativity and even more the new third stage of the theory, the unitary field theory, from other physical theories are the degree of formal speculation, the slender empirical basis, the boldness in theoretical construction and, finally, the fundamental reliance on the uniformity of the secrets of natural law and their accessibility to the speculative intellect. It is this feature which appears as a weakness to physicists who incline toward realism or positivism, but is especially attractive, nay, fascinating, to the speculative mathematical mind. [Emile] Meyerson in his brilliant studies on the theory of knowledge justly draws a comparison of the intellectual attitude of the relativity theoretician with that of Descartes, or even of Hegel, without thereby implying the censure which a physicist would read into this.

However that may be, in the end experience is the only competent judge.

Yet in the meantime one thing may be said in defense of the theory. Advance in scientific knowledge must bring about the result that an increase in formal simplicity can only be won at the cost of an increased distance or gap between the fundamental hypothesis of the theory on the one hand and

the directly observed facts on the other hand. Theory is com-
pelled to pass more and more from the inductive to the deductive method, even though the most important demand to
be made of every scientific theory will always remain: that it
must fit the facts.[3]

Apparently most of his fellow physicists concerned with atoms
and stars after 1930 were not convinced either that they understood
Einstein's latest enthusiasm or that Einstein fully understood their
own fields of endeavor. Einstein naturally prefered the Broglie-
Einstein-Schrödinger interpretation of atomic phenomena to the
Bohr-Heisenberg-Dirac formulation. And yet since Dirac had
bridged the gap between relativity and quantum mechanics, predicting in the process the positron, which then was promply discovered in different ways,[4] the consensus of informed opinion held
that Einstein's preferences must be merely stylistic, a matter of
taste, perhaps conditioned by his seemingly growing mathematical
mysticism. But Planck, Broglie, and Schrödinger, at least, tended to
agree with Einstein in later years that the real physical world
would ultimately demand a continuum mechanics.[5]

In 1936, Einstein published one of his most astute essays outlin-
ing his mature position on "Physics and Reality." Two paragraphs
from that work show especially well his attitudes toward the new
quantum theory and the fundamentals of physics:

Probably never before has a theory been evolved which has
given a key to the interpretation and calculation of such a
heterogeneous group of phenomena of experience as has the
quantum theory. In spite of this, however, I believe that the
theory is apt to beguile us into error in our search for a uniform basis for physics, because, in my belief it is an *incomplete*
representation of real things, although it is the only one
which can be built out of the fundamental concepts of force
and material points (quantum corrections to classical
mechanics). The incompleteness of the representation is the
outcome of the statistical nature (incompleteness) of the laws.
I will now justify this opinion.

Then after attempting to encourage others not to rest content with probabilistic and statistical interpretations merely because they were so pragmatic and fruitful for the present, Einstein returned to what he perceived as a real threat:

> To be sure, it has been pointed out that the introduction of a space-time continuum may be considered as contrary to nature in view of the molecular structure of everything which happens on a small scale. It is maintained that perhaps the success of the Heisenberg method points to a purely algebraical method of description of nature, that is to the elimination of continuous functions from physics. Then, however, we must also give up, by principle, the space-time continuum. It is not unimaginable that human ingenuity will some day find methods which will make it possible to proceed along such a path. At the present time, however, such a program looks like an attempt to breathe in empty space.[6]

When those words were written Einstein was fifty-seven years old, past the peak of his scientific productivity but in the midst of his busiest writing period. Having resigned his three positions in Berlin right after the Nazi regime came to power in 1933, and having decided to settle at Princeton rather than in southern California, Einstein was living quietly and working daily at the Institute for Advanced Study. But he was at the height of his fame and in great demand as a guru or promoter for all good causes. Consequently he shifted more and more from physics toward philosophy.

As shown in Figures III and IV on the next page, Einstein achieved a lifetime record of about 274 separate scientific papers and published notes. He was a consistent, though not especially prolific, contributor to professional publications who produced technical papers in spurts appropriate to what he had to say. Compared against his professional lifetime average (54 years divided into the 274-paper corpus) of more than 5 publications per year, Einstein's later years were by no means barren (more than 2 papers per year), but his general writings of all sorts came to outnumber his scientific works by about 35.

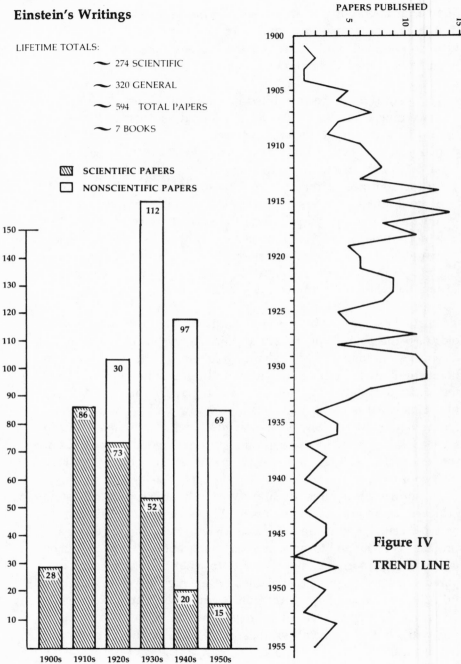

Figure III

Einstein's Writings

LIFETIME TOTALS:

— 274 SCIENTIFIC

— 320 GENERAL

— 594 TOTAL PAPERS

— 7 BOOKS

▨ **SCIENTIFIC PAPERS**
☐ **NONSCIENTIFIC PAPERS**

PAPERS PUBLISHED

Figure IV

TREND LINE

SOURCES: Nell Boni *et al. A Bibliographical Checklist . . . to the Published . . . of Albert Einstein* (1960); cf. Ernst Weil, Margaret C. Shields, Martin J. Klein, and Nandor L. Balasz in *Dictionary of Scientific Biography.*

Although only 30 or so of his technical publications were of major importance and fewer than half of those were of sufficient length and breadth to be quickly recognized by his peers, still it should be noted that the scientific Einstein generally worked and published alone. Only thirty-six of the items listed among his scientific articles show credit shared with collaborators. Half of those shared papers were written after his emigration to the United States. In all, they show only twenty-two different names, with eight appearing more than once. Only four papers were authored by Einstein with *two* collaborators. Twelve collaborators had the privilege of doing more than one paper with him. The Viennese mathematician Walter M. Mayer, who was Einstein's traveling companion during the trans-Atlantic trips of 1930 to 1933, co-authored at least seven articles relating to vector and spinor analysis and the intricacies of unified field theory. All told, Einstein's published work with collaborators was a minor though significant fraction (about one-eighth) of his total output. Enough has been learned of his life, working habits, and communications channels, however, to refute the legend that he was a solitary genius.[7]

Einstein often relished the chance to say that he could work alone and anywhere, that the life of a lighthouse-keeper should be ideal for a theoretical physicist, and that work as a cobbler or plumber or at some other craft while doing physics ought to prove salutary to the science as well as to the worker. But clearly his acknowledged debts to friends, colleagues alive and dead, students, and correspondents were very great indeed.

One might even argue that he was too thoroughly socialized and specialized in his first two decades of professional life (1909–29) to communicate effectively with the growing number of scientists outside of mathematical physics. Although fellow specialists appreciated him readily enough, few understood him well, because of his radical challenges to conventional assumptions. His most devoted disciples tended to mistake or misjudge Einstein's own inferences from relativity theory. As he evolved from his quest for electrodynamic invariances through his quest for the unified field, Einstein progressed from an essentially positivistic world-view to an essentially rationalistic one. This personal philosophical pilgrimage was seldom appreciated by his followers. They wanted him to be what he had been, not what he was or would become.

Despite many biographies, Einstein's life and thoughts remain elusive. He was a contemplative man with a good sense of humor. He played many roles within an elite subculture that valued physical intuition and mathematical insight above the experimentalism of most physical scientists. Einstein sometimes sacrificed his sense of community for the hopes of inspiration. So many of his colleagues and peers rendered him such reverence as the apostle of relativity that he became a celebrity as well as an authority. He was expected to do revelations as well as research.

This dual status, achieved too soon for the healthy consideration of his later life's work, undoubtedly affected the reception of his offerings on the unitary field in 1929 and 1949. Fierce professional allegiances complicated by nationalistic loyalties during and after World War I had led to the apotheosis of Einstein on the left and his anathema on the right. Theoretical physics was, it is true, attractive to Jewish intellectuals, and relativity theory was attractive to radical revolutionaries of many persuasions, but the turmoil of the Weimar Republic was breeding extremism and polarization without respect to character. Einstein simply could never deny his private convictions in favor of his public image.[8]

Rather than relativity theory, special or general, it was Einstein's declared metaphysical quest for a unified physics and philosophy, embodied both in his personal history (for example, the "Manifesto to Europeans") and in his epistemological stance (vis-à-vis Karl Marx, Ernst Mach, Henri Bergson, and others) that did the most to make him revered in later life. He suffered much personal loss at the hands of the Nazis, and he fought so hard for social justice, economic equity, and the demise of the nation-state that he was in spite of himself either loved or hated.[9]

It was clear to men everywhere, even those professing not to understand a word of his science, that Einstein was not merely a man of science but a lover of wisdom. His dedication to truth, justice, and beauty had long before his death made his name a talisman, a symbol of genius. And so we must inquire a few steps farther into the private values of Einstein the public figure. We must ask how his philosophy of nature was related to his religious attitudes, not in an institutional sense but in an individual sense.[10]

If a man's religion is defined as his ultimate concern with whatever values he honors most profoundly, then no one is irreligious. Although cultural assumptions, inherited creeds, and institutional churches wreak havoc with our efforts to understand this fact, I believe that the alpha and omega of the humanities, especially when applied to the study of great leaders of imaginative movements among men, are religion. To say that nature, science, or physics was Einstein's religion is not enough to satisfy our wonder. How did he confront the ultimate mysteries of our consciences and consciousness?

His first foray into the forum of psychical as opposed to physical philosophy had happened in London on 27 October 1930, when George Bernard Shaw reverently introduced Einstein to speak on the subject of "Cosmic Religion." "Everything that men do or think," began Einstein on that occasion, "concerns the satisfaction of the needs they feel or the escape from pain." Religion has, he argued, three instinctive sources: it grows out of fear, or out of social feelings, or out of a "cosmic religious sense." This last source characterizes the symbolic roles played by the Buddha, Moses, David, the Christ, and even Schopenhauer. "The religious geniuses of all times have been distinguished by this cosmic religious sense, which recognizes neither dogmas nor God made in man's image." Democritus, Francis of Assisi, and Spinoza served as his exhibits. The most important function of art and science is to arouse and nurture this feeling of cosmic identity, although ethics and morality can be supported by the religious source of social solidarity alone. In conclusion Einstein said:

> I assert that the cosmic religious experience is the strongest and noblest driving force behind scientific research. . . . Only those who have dedicated their lives to similar ends can have a living conception of the inspiration which gave these men the power to remain loyal to their purpose in spite of countless failures. It is the cosmic religious sense which grants this power.[11]

Shaw may have regretted his overenthusiastic introduction upon hearing Einstein end by repeating the absurd claim that "the only deeply religious men of our materialistic age are earnest scientists."

In 1911, Einstein the atheist had been forced by the mores of the Austro-Hungarian Empire to declare a religious preference when accepting the post at the German University of Prague: "Mosaic creed" had been his answer for that record. But by 1921, when he first visited the United States, Einstein the agnostic, in company with Chaim Weizmann on behalf of Zionism, was confronted by a New York rabbi who cabled him for a direct response to the question "Do you believe in God?" Einstein's oracular answer still rings as the truest capsule of his credo: "I believe in Spinoza's God, who reveals himself in the orderly harmony of what exists; not in a God who concerns himself with the fates and actions of human beings."[12] Wiseacres who might consider this thirty-word creed the cleverest of circumlocutions could not be expected to have read Spinoza. Those wise enough to have considered the parallels in the lives and thoughts of Baruch Spinoza and Albert Einstein could not fail to see the former as an ever ready source of inspiration to the latter. Both were displaced rationalist philosophers who fought dogmatic orthodoxies with logic and compassion.

Similar kinds of logical passion and passionate logic manifested themselves in other persons also during these decades. Bertrand Russell and Alfred North Whitehead, for instance, developed their own strains of symbolic logic to synthesize mathematics with language. Their *Principia Mathematica* held its own fascination, parallel to Einstein's relativity theories and Freud's psychoanalytic theories, for a whole generation of avante-garde philosophers. The "Vienna Circle" of positivists, beginning with Moritz Schlick and ending, perhaps, with Hans Reichenbach, were all profoundly impressed by relativity, but they were hardly aware of Einstein's philosophic pilgrimage. In fact, they often discounted history for their philosophy. Even more recently, groups as diverse as the Neo-Thomists of Roman Catholicism and the Marxist-Leninists of Soviet Russia have found Einstein's reputation, if not his character, a philosophic landmark. Einstein's religious views were often buried amid the pleas of special-interest groups.[13]

In 1929, Einstein the pundit wrote for the anthology *Living Philosophers* a sketch that, though not published until 1931, vividly amplified his stance as a spokesman for the realists among natural philosophers:

> The most beautiful thing we can experience is the mysterious. It is the source of all true art and science. He to whom this emotion is a stranger, who can no longer pause to wonder and stand rapt in awe, is as good as dead: his eyes are closed. This insight into the mystery of life, coupled though it be with fear, has also given rise to religion. To know that what is impenetrable to us really exists, manifesting itself as the highest wisdom and the most radiant beauty which our dull faculties can comprehend only in their most primitive forms—this knowledge, this feeling, is at the center of true religiousness. In this sense, and in this sense only, I belong in the ranks of devoutly religious men.
>
> I cannot imagine a God who rewards and punishes the objects of his creation, whose purposes are modeled after our own—a God, in short, who is but a reflection of human frailty. Neither can I believe that the individual survives the death of his body, although feeble souls harbor such thoughts through fear or ridiculous egotism. It is enough for me to contemplate the mystery of conscious life perpetuating itself through all eternity, to reflect upon the marvelous structure of the universe which we can dimly perceive, and to try humbly to comprehend even an infinitesimal part of the intelligence manifested in nature.[14]

Only a decade later Einstein the militant anti-Nazi marveled how that passage sounded "true as ever" yet "curiously remote and strange." "How can that be?" he asked in 1939. "Has the world changed so profoundly in ten years, or is it merely that I have grown ten years older? . . . What are ten years in the history of humanity? . . . Is my critical reason so susceptible that the physiological change in my body during those ten years has been able to influence my concept of life so deeply?" Hitler's Germany had intervened, and Einstein's faith was being tested.

"No, something quite different is involved," he continued. "In these ten years confidence in the stability, yes, even the very basis for existence, of human society has largely vanished. One senses not only a threat to man's cultural heritage, but also that a lower value is placed upon all that one would like to see defended at all costs."[15]

This public position of "practical atheism" or *a*-theism (any deist is by definition not a theist) led Einstein into innumerable minor difficulties as a celebrity. He was accused of being a pantheist, a naïve realist, and a "God-intoxicated man," just as Spinoza had been accused in the seventeenth century.

None of these embarrassments was so profoundly troubling, however, as the problem of reconciling his public pacifism with his fear that the Nazi megamachine might develop nuclear-fission weapons. In 1932, Einstein and Freud as internationalists had exchanged lengthy open letters that the League of Nations publicized in a multilingual booklet entitled *Why War?* Only seven years later, it was clear that uranium in German laboratories had been changed to barium after bombardment by neutrons. Thus the simplified formula, $E = mc^2$ from 1905 now appeared ominously close to spectacular and possibly horrendous demonstration. With this intelligence communicated by Bohr, Einstein knew that he must now, in the name of humanity, sign another letter, this one very private to Franklin Delano Roosevelt. He did so with fear and trembling on 2 August 1939. After the bombs of 6 and 9 August 1945, he felt the sin of Hiroshima and Nagasaki: "Yes, I pressed the button," he admitted some years later to one of his biographers.[16]

With those, like Emerson, who say that "consistency is the hobgoblin of small minds" Einstein the pacifist could now sympathize. But Einstein the physicist could never agree to this compromise with his fundamental epistemological principles. He, like J. Robert Oppenheimer, James Franck, Leo Szilard, and others, learned to know the meaning of Greek tragedy in its most acute private form during the late 1940s and early 1950s. The irony that pure and proud scientists should come to share the humiliating knowledge of good and evil, incarnated in human political dilemmas, was a challenge to which they (and we) have yet fully to respond.

Like Albert Schweitzer, Bertrand Russell, Rabindranath Tagore,

Mahatma Gandhi, and other elder statesmen of the intellect, all of whom he admired and with whom he corresponded, Einstein responded in later life by collecting, polishing, and publishing his many short essays on moral, social, and natural philosophy. His last two anthologies, *Out of My Later Years* (1950) and *Ideas and Opinions* (1954), built upon his first collection of such essays from 1934, *Mein Weltbild (The World as I See It)*. Einstein the socialist dared to reveal himself and his hopes for world government, effective United Nations peace-keeping machinery, and a commonwealth of man. His words on religion and science must be studied in conjunction with the words of Maimonides, Spinoza, and perhaps Whitehead and Russell to savor fully his search and solution:

> I have now reached the point [Einstein wrote in 1949] where I may indicate briefly what to me constitutes the essence of the crisis of our time. It concerns the relationship of the individual to society. The individual has become more conscious than ever of his dependence upon society. But he does not experience this dependence as a positive asset, as an organic tie, as a protective force, but rather as a threat to his natural rights, or even to his economic existence. Moreover, his position in society is such that the egotistical drives of his make-up are constantly being accentuated, while his social drives, which are by nature weaker, progressively deteriorate. All human beings, whatever their position in society, are suffering from this process of deterioration. Unknowingly prisoners of their own egotism, they feel insecure, lonely, and deprived of the naive, simple, and unsophisticated enjoyment of life. Man can find meaning in life, short and perilous as it is, only through devoting himself to society.[17]

Even this mild form of socialist commitment brought on the enmity of some of his less sophisticated colleagues in the United States. But Einstein, despite the worldwide depression, was no admirer of the Soviet system under Stalin. Rather, his faith in democracy grew as his hopes for a compromised peace waned. One of his most beautiful sentences passed judgment on "Why science?": "It stands to the everlasting credit of science that by acting on the human mind it has overcome man's insecurity before himself and before nature." And again he wrote in regard to "Why science education?": "The general public may be able to follow the details of scientific research to only a modest degree; but it can register at least one great and important gain: confidence that human thought is dependable and natural law universal."[18]

After he became a naturalized U.S. citizen in 1940, Einstein was feted and lionized as America's most illustrious immigrant. While war clouds boiled over Europe and Asia in 1941, Einstein spoke with serene feeling before a conference in New York on "Science, Philosophy and Religion in Their Relation to the Democractic Way of Life." His definitions of science and religion, similar to those of Hume and Kant, have since become classics:

Science is the century-old endeavor to bring together by means of systematic thought the perceptible phenomena of this world into as thoroughgoing an association as possible. To put it boldly, it is the attempt at the posterior reconstruction of existence by the process of conceptualization. But when asking myself what religion is I cannot think of the answer so easily. . .[19]

. . . instead of asking what religion is I should prefer to ask what characterizes the aspirations of a person who gives me the impression of being religious: A person who is religiously enlightened appears to me to be one who has, to the best of his ability, liberated himself from the fetters of his selfish desires and is preoccupied with thoughts, feelings, and aspirations to which he clings because of their super-personal value. It seems to me that what is important is the force of this super-personal content and the depth of the conviction concerning its overpowering meaningfulness regardless of whether any attempt is made to unite this content with a divine Being, for otherwise it would not be possible to count Buddha and Spinoza as religious personalities. Accordingly, a religious person is devout in the sense that he has no doubt of the significance and loftiness of those super-personal objects and goals which neither require nor are capable of rational foundation. They exist with the same necessity and matter-of-factness as he himself. In this sense religion is the age-old endeavor of mankind to become clearly and completely conscious of these values and goals and constantly to strengthen and extend their effect. If one conceives of religion and science according to these definitions, then a conflict between them appears impossible. For science can only ascertain what *is*, but not what *should be*, and outside of its domain value judgments of all kinds remain necessary. Religion, on the other hand, deals only with evaluations of human thought and action: it cannot justifiably speak of facts and relationships between facts. . . .[20]

Now, even though the realms of religion and science in themselves are clearly marked off from each other, nevertheless there exist between the two strong reciprocal relationships and dependencies. Though religion may be that which determines the goal, it has, nevertheless, learned from science, in the broadest sense, what means will contribute to the attainment of the goals it has set up. But science can only be created by those who are thoroughly imbued with the aspiration toward truth and understanding. This source of feeling, however, springs from the sphere of religion. To this there also belongs the faith in the possibility that the regulations valid for the world of existence are rational, that is, comprehensible to reason. I cannot conceive of a genuine scientist without that profound faith. The situation may be expressed by an image: Science without religion is lame, religion without science is blind.[21]

Einstein's deterministic, materialistic philosophy depended heavily on Spinoza's distinctions between determined means and freely chosen human ends *and* on the etherealized "matter" of modern physics. Still, his professional argument with Niels Bohr and the majority of quantum mechanicians over the permanent value of probabilistic attitudes in physics continued until his death on 18 April 1955. Einstein the epistemologist delighted in provoking these pragmatic statisticians by invoking the name of Spinoza's God (read "Nature") wherever he could. "I cannot believe," he often said, "that God plays dice with the Universe." The most famous of these sayings was carved above the fireplace in the professors' lounge of the old mathematics building at Princeton: RAFFINIERT IST DER HERR GOTT, ABER BOSHAFT IST ER NICHT (God is subtle, but he is not malicious). This aphorism, above all others uttered by Albert Einstein, expresses most succinctly his profound faith in rationalism, the concord we presume between nature and reason's ability to know it.[22] It is an article of the rationalist's faith that nature may be ever so intricate and yet never so contrary as to thwart man's efforts to comprehend it. This article may be said to be the "golden rule" of Greco-European science and Judeo-Christian-Muslim philosophy, as opposed to theology. It is not, however, the only rule by which science and philosophy have moved onward. Only a wider grasp of world history and of comparative metaphysics than Einstein possessed can show us these alternatives.[23]

By defining "God" as nature, "science" as philosophy, and "religion" as the quest for consciousness *and* control over values, Einstein was clear, even if most of us are not, in his meaning for the words "Science without religion is lame, religion without science is blind." His highest synthesis is caught by this couplet. That philosophy without values cannot run and that values without philosophy cannot see are two premises basic to Einstein's faith in progress and in vision.

NOTES

1. See, e.g., Arthur F. Bentley, *Relativity in Man and Society* (New York: Putnams, 1926), pp. 6–14, 229–32. Helmut Schoek and James W. Wiggins, eds., *Relativism and the Study of Man* (Princeton: Van Nostrand, 1961). See also Harry Woolf, ed., *Quantification: A History of the Meaning of Measurement in the Natural and Social Sciences* (Indianapolis: Bobbs-Merrill, 1961). W. H. McCrea, "Cosmology after Half a Century," *Science*, 160 (21 June 1968): 1295–99.
2. See, e.g., one of the standard histories of anthropology, such as Robert H. Lowie, *The History of Ethnological Theory* (New York: Farrar-Rinehart, 1937); or A. L. Kroeber, ed., *Anthropology Today: An Encyclopedic Inventory* (Chicago: Univ. of Chicago Press, 1953). Cf. Michael

Polanyi, *Personal Knowledge: Toward a Post-Critical Philosophy* (Chicago: Univ. of Chicago Press, 1958). Harper Torchbook edition, New York, 1962; and John Ziman, *Public Knowledge: An Essay concerning the Social Dimension of Science* (Cambridge: Univ. Press, 1968).

3. A. Einstein, "Field Theories, Old and New," *New York Times*, 3 February 1929, and in pamphlet form (New York: Readex Microprint Corp., 1960), pp. 3–4. See also Emile Meyerson, *Identity and Reality*, trans. Kate Loewenberg (New York: Dover, 1962), orig. edn. 1908; and *La Déduction relativiste* (Paris: Payot, 1925).

4. Norwood Russell Hanson, *The Concept of the Positron: A Philosophical Analysis* (Cambridge: Univ. Press, 1963), esp. pp. 135–65.

5. Max Planck, *Where is Science Going?* trans. James Murphy (New York: Norton, 1932), pp. 195–200. Erwin Schrödinger, *Mind and Matter* (Cambridge: Univ. Press, 1958). Louis de Broglie, *The Current Interpretation of Wave Mechanics: A Critical Study* (Amsterdam: Elsevier, 1964).

6. A. Einstein, "Physics and Reality," *Journal Franklin Institute*, Vol. 221, no. 3 (March 1936), trans. Jean Piccard; conveniently available as item 13 in Einstein's *Out of My Later Years*, rev. edn. (Westport, Conn.: Greenwood Press, 1970), pp., 88, 92.

7. Nell Boni, Monique Russ, and Don H. Laurence, comps., *A Bibliographical Checklist and Index to the Published Writings of Albert Einstein* (New York: Readex Microprint, 1960). Ernest Weil, *Albert Einstein . . . a Bibliography of His Scientific Papers: 1901–1954* (London: Goldschmidt, 1960). Margaret C. Shields, Bibliographical Appendix to P. A. Schilpp, ed., *Albert Einstein: Philosopher-Scientist*, 2 vols. (New York: Harper Torchbooks, 1959), vol. II, pp. 691–760.

8. See Fritz Ringer, *The Decline of the German Mandarins: The German Academic Community, 1890–1933* (Cambridge: Harvard Univ. Press, 1969); Joseph Haberer, *Politics and the Community of Science* (New York: Van Nostrand, Reinhold, 1969); Paul Forman, "Scientific Internationalism and the Weimar Physicists," *ISIS* 64 (June 1973): 151–80. Stanley Coben, "The Scientific Establishment and the Transmission of Quantum Mechanics to the United States, 1919–1932," *American Historical Review* 76 (April 1971): 442–66.

9. See, e.g., Boris Kuznetsov, *Einstein*, trans. V. Talmy (Moscow: Progress Publisher, 1965); John T. Blackmore, *Ernst Mach: His Life, Work and Influence* (Berkeley: Univ. of California Press, 1972); P. A. Y. Gunter, ed. and trans., *Bergson and the Evolution of Physics* (Knoxville: Univ. of Tennessee Press, 1969); Herbert L. Samuel, *Essay in Physics . . .* (New York: Harcourt Brace, 1952); cf. Hans Reichenbach, *Axiomatization of the Theory of Relativity* (Berkeley: Univ. of California Press, 1969); Mendel Sachs, *Ideas of the Theory of Relativity: General Implications from Physics to Problems of Society* (Jerusalem: Israel Univ. Press, 1974).

10. For background to what follows, see Edwin A. Burtt, *Man Seeks the Divine: A Study in the History of Comparative Religions*, 2nd edn. (New

York: Harper & Row, 1970); and cf. Bertrand Russell, *Why I Am Not a Christian; and Other Essays on Religion . . .*, ed. Paul Edwards (New York: Simon and Schuster, 1957).

11. Albert Einstein, *Cosmic Religion: with Other Opinions and Aphorisms* (New York: Covici-Friede, 1931), pp. 1, 49, 52.

12. See Banesh Hoffmann with Helen Dukas, *Albert Einstein: Creator and Rebel* (New York: Viking Press, 1972), p. 95. Chapman Cohen, *God and the Universe* (London: Pioneer Press for the Secular Society, 1946), 3rd edn., pp. 127–28.

13. For some indication of the breadth of Einstein's influence on diverse imaginations, see P. W. Bridgman, *A Sophisticate's Primer of Relativity* (Middletown, Conn.: Wesleyan Univ. Press, 1962); John F. Kiley, *Einstein and Aquinus: A Rapprochement* (The Hague: M. Nijhoff, 1969); Boris Kuznetsov, *Einstein and Dostoyevsky*, trans. V. Talmy (London: Hutchinson International, 1972).

14. A. Einstein, essay in *Living Philosophies* (New York: Simon and Schuster, 1931), pp. 6–7; cf. Joseph Needham, ed., *Science, Religion and Reality* (New York: Braziller, 1955), orig. edn. London, 1925. See also Michael Polanyi, *Science, Faith and Society* (Chicago: Univ. of Chicago Press, 1964), orig. edn. (London: Oxford University Press, 1946.

15. Einstein in *I Believe*, ed. Clifton Fadiman (New York: Simon and Schuster, 1939), p. 367; cf. A. Einstein [from *Science* 91 (24 May 1940): 487–92] in W. Warren Wagar, ed., *Science, Faith, and Man: European Thought since 1914* (New York: Harper Torchbooks, 1968).

16. Einstein to Antonina Valentin, *Einstein: A Biography*, p. 271; but cf., Einstein *Ideas and Opinions* (New York: Bonanza, 1954), p. 121. See also Albert Einstein–Sigmund Freud, *Why War* [2 letters from 1932] (Paris: International Institute of Intellectual Cooperation, 1933); cf. David Irving, *The German Atomic Bomb: The History of Nuclear Research in Nazi Germany* (New York: Simon and Schuster, 1967).

17. Einstein, *Out of My Later Years* (New York: Philosophical Library, 1950), p. 123, from *Monthly Review* (New York, 1949). Cf. Adrienne Koch, ed., *Philosophy for a Time of Crisis* (New York: Dutton, 1960), p. 96.

18. Einstein, "Science and Society," no. 20 in *Out of My Later Years* (Greenwood edn.), p. 137, orig. essay from 1935–1936, trans. H. and R. Norden.

19. Einstein, *Out of My Later Years*, p. 24.

20. Ibid., p. 25.

21. Ibid., p. 26; also, "Ideas and Opinions," pp. 44–46, from *Science, Philosophy and Religion* (New York: Conference on Science, Philosophy and Religion in Relation to the Democratic Way of Life, 1941). See also Irving Kristol, "Einstein: The Passion of Pure Reason," in *Perspectives U.S.A.*, no. 14 (Winter 1956): pp. 76–91.

22. See Hoffmann with Dukas, p. 146. This famous aphorism was uttered in response to a discussion about Dayton C. Miller's experimental

challenge in the late 1920s of the usual interpretation of the Michelson-Morley aether-drift experiment.

23. See, for example, Fritjof Capra, *The TAO of Physics: An Exploration of the Parallels Between Modern Physics and Eastern Mysticism* (Boulder, Colo.: Shambhala, 1975), or Ronald Duncan and Miranda Weston-Smith, eds., *The Encyclopaedia of Ignorance* (Elmsford, N.Y.: Pergamon Press, 1977).

Bibliographical Note

The notes that accompany the chapters in this book are fairly complete in guiding the serious student of Einstein and relativity to the primary and secondary sources currently available in English. But there is more, much more, tertiary material than need be indicated here. And in time much more primary material will become available as correspondence, memoirs, and historical monographs are published. The Einstein Archives at Princeton are in the process of being cataloged. The Niels Bohr Archives at the American Institute of Physics in New York City, meanwhile, remain the central repository for the serious study of the history and philosophy of the physical sciences in the twentieth century.

The "Note on Sources" and "Bibliography" in my book *The Ethereal Aether: A History of the Michelson-Morley-Miller Aether-Drift Experiments, 1880–1930* (Austin: Univ. of Texas Press, 1972) remain, so far as I know, the best initial guide to the secondary literature on the subjects of this present work as well. Almost 650 books relevant to the genesis of relativity are listed there. But as Gerald Holton has pointed out, the literature on relativity theory is much vaster than that. (See his essay "Origins of the Special Theory of Relativity" in *Thematic Origins of Scientific Thought: Kepler to Einstein* [Cambridge: Harvard Univ. Press, 1973], p. 166.)

Furthermore, the problem of audience that every author faces is complicated immeasurably by this, one of the most complex of subjects. Each student must ultimately thread his own way through such a jungle and find his own path to understanding. But the core of the technical content may best be approached through such works as Peter G. Bergmann's *Introduction to the Theory of Relativity* (Englewood Cliffs: Prentice-Hall, 1942) and John Lighton Synge's *Relativity: The Special Theory* (Amsterdam: North Holland, 1956) and *Relativity: The General Theory* (New York: Interscience, 1960).

There are five major bibliographies and checklists that give annotated introductions to Einstein's writings. In chronological order for comparison, they are as follows:

1. Ernest Weil. *Albert Einstein: A Bibliography of His Scientific Papers*, 1901–30 (London: Goldschmidt, 1937), rev. ed., 1901–54 (1960).
2. Margaret C. Shields. "Bibliography of the Writings of Albert Einstein to May 1951," in P. A. Schilpp, ed., *Albert Einstein: Philosopher-Scientist* (New York: Harper Bros., 1951), vol. II, pp. 691–760.
3. Nell Boni, Monique Russ, and Dan H. Laurence, eds. *A Bibliographical Checklist and Index to the Published Writings of Albert Einstein* (Patterson, N.J.: Pageant Books, 1960).
4. Herbert S. Klickstein. "A Cumulative Review of Bibliographies of the Published Writings by Albert Einstein," *Journal of Albert Einstein Medical Center* 10, 3 (July 1962), 141–54.
5. J. T. Combridge, ed. *Bibliography of Relativity and Gravitation Theory, 1921 to 1937* (London: Kings College Press, 1965).

In addition, ongoing compilations by scholars such as André Mercier in Berne and Alvin E. Jaeggli in Zurich complement the work of others cataloguing materials at the Einstein Archives in Princeton, the Bohr Library in New York, and wherever else Einsteiniana is held.

For historical purposes, I suggest here a short list of books in English that seem particularly valuable as aids to further study of the origins of Einstein's theories of relativity:

Bernstein, Jeremy. *Einstein* (New York: Viking Press, 1973).
Born, Max. *Einstein's Theory of Relativity*, rev. ed. (New York: Dover, 1962).
Clark, Ronald W. *Einstein: The Life and Times* (New York: World Publishing Co., 1971).
Einstein, Albert, and Leopold Infeld. *The Evolution of Physics* (New York: Simon & Schuster, 1938).
Frank, Philipp. *Einstein: His Life and Times* (New York: Alfred A. Knopf, 1967).
Hoffmann, Banesh, with Helen Dukas. *Albert Einstein: Creator and Rebel* (New York: Viking Press, 1972).

Lorentz, H. A. *et al. The Principle of Relativity: A Collection of Original Memoirs . . .*, trans. W. Perrett and G. B. Jeffery (New York: Dover, 1923).

North, John D. *The Measure of the Universe: A History of Modern Cosmology* (Oxford: Clarendon Press, 1965).

Planck, Max. *Scientific Autobiography and Other Papers*, trans. F. Gaynor (New York: Philosophical Library, 1949).

Schilpp, Paul A., ed. *Albert Einstein: Philosopher-Scientist*, 2 vols. (New York: Harper & Row, 1959).

Seelig, Carl. *Albert Einstein: A Documentary Biography*, trans. M. Savill (London: Staples, 1956).

Weinberg, Steven. *Gravitation and Cosmology* (New York: John Wiley & Sons, 1972).

Wheeler, John A., Kip S. Thorne, and B. Kent Harrison. *Gravitation Theory and Gravitational Collapse* (Chicago: University of Chicago Press, 1965).

Williams, L. Pearce, ed. *Relativity Theory: Its Origins and Impact on Modern Thought* (New York: John Wiley & Sons, 1968).

Index

259